Agricultural Microbiology

Dr. Anoop Singh

RANDOM

RANDOM PUBLICATIONS

NEW DELHI (INDIA)

Agricultural Microbiology

ISBN 978-93-5111-322-5

Published in 2014 in India by

RANDOM PUBLICATIONS

4376-A/4B, Gali Murari Lal, Ansari Road
New Delhi-110 002
Phone : +91-11-43580356, +91-11-23289044
e-mail: randomexports@gmail.com, sales@randompublications.com, info@randompublications.com

Reprinted 2023

Type Setting by : Keystoneprintads, Delhi-110051
Printed at : Replika Press Pvt. Ltd.

Preface

Agricultural microbiology is a branch of microbiology dealing with plant-associated microbes and plant and animal diseases. It also deals with the microbiology of soil fertility, such as microbial degradation of organic matter and soil nutrient transformations.

Outcrossing of transgenic plants not only poses potential environmental risks, it may also trouble farmers and food producers. Many countries have different legislations for transgenic and conventional plants as well as the derived food and feed, and consumers demand the freedom of choice to buy GM-derived or conventional products. Therefore, farmers and producers must separate both production chains. This requires coexistence measures on the field level as well as traceability measures throughout the whole food and feed processing chain. Research projects such as Co-Extra, SIGMEA and Transcontainer investigate how farmers can avoid outcrossing and mixing of transgenic and non-transgenic crops, and how processors can ensure and verify the separation of both production chains.

In the biosphere, microorganisms play a key role in maintaining the chemical balance of available nutrients and metabolic waste products. In this way, they help preserve the natural environment. In addition, microorganisms are important in the elemental cycles occurring in the soil.

An ecosystem is the community of organisms found in a physically defined space. In the soil ecosystem, microorganisms are the largest contributors of organic matter. This organic matter is derived from the metabolism of animal and plant waste. The decay-resistant organic matter not recycled combines with mineral particles to form the dark-coloured material of soil called humus. Microbiology is a broad term which includes virology, mycology, parasitology, bacteriology and other branches.

A microbiologist is a specialist in microbiology and these other topics. Microbiology is researched actively, and the field is advancing continually. It is estimated only about one per cent of all of the microbe species on Earth have been studied. Although microbes were directly observed over three hundred years ago, the field of microbiology can be said to be in its infancy relative to older biological disciplines such as zoology and botany. This book will be useful to researchers, teachers and students of microbiology, agriculture, botany and zoology.

I thank all associates of my group who have helped in the preparation of the book. My individual thanks go to "Random Publications" who have in print the book.

— Dr. Anoop Singh

Contents

1

Microbes and Agriculture

Microbes and Agriculture - Besides being important in biogeochemical cycling of nutrients, microbes play vital role in maintenance of soil fertility and in crop protection.

SOIL FERTILITY

Soil Fertility - Microbes are being exploited in two important ways - biofertilisers, and creating new nitrogen-fixing organisms.

Biofertilisers

Potential of Rhizobium, Azotobacter, Beijerinckia, Azospirillum, Cyanobacteria, such as species of Aulosira, Anabaena, Nostoc, Plectonema, Scytonema, Tolypothrix, and Azalia as biofertilisers has been exploited so as these could serve an alternative to chemical fertilisers.

Many brands of rhizobial inoculants are already in market today in the country. Several organisation and manufacturers are producing huge quantities of Rhizobium culture in the country.

These include Micro Bac., India, Shyam nagar, Parganas; Bacifil Inoculants, Lucknow; Govt. of Tamil Nadu; Nitro Fix Industries, Calcutta (W. Bengal) and Indian Organic Chemical Ltd. Bombay. In some other States, units are being prepared for increase in its production. Much progress has also been made with cyanobacteria in this direction.

Mycorrhizae, both ecto and endomycorrhiza help in uptake of N, P, K and Ca. They, particularly help in phosphorous nutrition.

Nitrogen Fixers

Nitrogen Fixers - Through recombinant DNA technology efforts have been made to introduce nitrogen-fixing genes (nif genes) into wheat, corn, rice, etc. Plasmids of the bacterium, E. coli and yeast are being worked out for such a possibility. Hybrid E. coli plasmid cloned with nif genes of a nitrogen-fixing bacterium, Klebsiella pneumoniae and hybrid yeast plasmids are then integrated.

Biopesticides

Biopesticides - Several microbes (viruses bacteria, and fungi) are being developed as suitable biopesticides for management of insect and nematodal pests. Some fungi have good potential of their use as bionematicides to control nematodal pests of vegetables, fruit and cereal crops. Some bacterial and fungal products are also in use to control diseases of roots and shoots of plants.

Bioweedicides

Bioweedicides - Several fungi have been found very useful in the control of troublesome weeds of crop fields. Registered products are available in market for use in several countries.

MICROBIOLOGY

STRUCTURE

Viruses display a wide diversity of shapes and sizes, called *morphologies*. Viruses are about 1/100th the size of bacteria. Most viruses that have been studied have a diameter between 10 and 300 nanometres. Some filoviruses have a total length of up to 1400 nm, however their diameters are only about 80 nm.

Most viruses are unable to be seen with a light microscope so scanning and transmission electron microscopes are used to visualise virus particles. To increase the contrast between viruses and the background, electron-dense "stains" are used. These are solutions of salts of heavy metals such as tungsten, that scatter the electrons from regions covered with the stain. When virus particles are coated with stain (positive staining), fine detail is obscured. Negative staining overcomes this problem by staining the background only.

A complete virus particle, known as a virion, consists of nucleic acid surrounded by a protective coat of protein called a capsid. These are formed from identical protein subunits called capsomers. Viruses can have a lipid "envelope" derived from the host cell membrane. The capsid is made from proteins encoded by the viral genome and its shape serves as the basis for morphological distinction. Virally coded protein subunits will self-assemble to form a capsid, generally requiring the presence of the virus genome. However, complex viruses code for proteins that assist in the construction of their capsid. Proteins associated with nucleic acid are known as nucleoproteins, and the association of viral capsid proteins with viral nucleic acid is called a nucleocapsid. The capsid and entire virus structure can be mechanically (physically) probed through atomic force microscopy. In general, there are four main morphological virus types:

Helical

These viruses are composed of a single type of capsomer stacked around a central axis to form a helical structure, which may have a central cavity, or hollow tube. This arrangement results in rod-shaped or filamentous virions: these can be short and highly rigid, or long and very flexible. The genetic material, generally single-stranded RNA, but ssDNA in some cases, is bound into the protein helix, by interactions between the negatively charged nucleic acid and positive charges on the protein. Overall, the length of a helical capsid is related to the length of the nucleic acid contained within it and the diameter is dependent on the size and arrangement of capsomers. The well-studied Tobacco mosaic virus is an example of a helical virus.

Icosahedral

Most animal viruses are icosahedral or near-spherical with icosahedral symmetry. A regular icosahedron is the optimum way of forming a closed shell from identical sub-units. The minimum number of identical capsomers required is twelve, each composed of five identical sub-units. Many viruses, such as rotavirus, have more than twelve capsomers and appear spherical but they retain this symmetry. Capsomers at the apices are surrounded by five other capsomers and are called pentons. Capsomers on the triangular faces are surround by six others and are call hexons.

Envelope

Some species of virus envelope themselves in a modified form of one of the cell membranes, either the outer membrane surrounding an infected host cell, or internal membranes such as nuclear membrane or endoplasmic reticulum, thus gaining an outer lipid bilayer known as a viral envelope. This membrane is studded with proteins coded for by the viral genome and host genome; the lipid membrane itself and any carbohydrates present originate entirely from the host. The influenza virus and HIV use this strategy. Most enveloped viruses are dependent on the envelope for their infectivity.

Complex

These viruses possess a capsid that is neither purely helical, nor purely icosahedral, and that may possess extra structures such as protein tails or a complex outer wall. Some bacteriophages, such as Enterobacteria phage T4 have a complex structure consisting of an icosahedral head bound to a helical tail, which may have a hexagonal base plate with protruding protein tail fibres. This tail structure acts like a molecular syringe, attaching to the bacterial host and then injecting the viral genome into the cell.

The poxviruses are large, complex viruses that have an unusual morphology. The viral genome is associated with proteins within a central disk structure known as a nucleoid. The nucleoid is surrounded by a membrane and two lateral bodies of unknown function. The virus has an outer envelope with a thick layer of protein studded over its surface. The whole particle is slightly pleiomorphic, ranging from ovoid to brick shape. Mimivirus is the largest known virus, with a capsid diameter of 400 nm. Protein filaments measuring 100 nm project from the surface. The capsid appears hexagonal under an electron microscope, therefore the capsid is probably icosahedral. Some viruses that infect Archaea have complex structures that are unrelated to any other form of virus, with a wide variety of unusual shapes, ranging from spindle-shaped structures, to viruses that resemble hooked rods, teardrops or even bottles. Other archaeal viruses resemble the tailed bacteriophages, although they can have multiple tail structures.

Genome

An enormous variety of genomic structures can be seen among viral species; as a group they contain more structural genomic diversity than plants, animals, archaea, or bacteria. A virus has either DNA or RNA genes and are called DNA viruses and RNA viruses respectively. By far most viruses have RNA.

Plant viruses tend to have single-stranded RNA and bacteriophages tend to have double-stranded DNA. Viral genomes are circular, such as polyomaviruses, or linear, such as adenoviruses. The type of nucleic acid is irrelevant to the shape of the genome. Among RNA viruses, the genome is often divided up into separate parts within the virion and is called *segmented*. Each segment often codes for one protein and they are usually found together in one capsid. Every segment is not required to be in the same virion for the overall virus to be infectious, as demonstrated by the brome mosaic virus.

A viral genome, irrespective of nucleic acid type, is either single-stranded or double-stranded. Single-stranded genomes consist of an unpaired nucleic acid, analogous to one-half of a ladder split down the middle. Double-stranded genomes consist of two complementary paired nucleic acids, analogous to a ladder. Some viruses, such as those belonging to the *Hepadnaviridae*, contain a genome that is partially double-stranded and partially single-stranded.

For viruses with RNA or single-stranded DNA, the strands are said to be either positive-sense (called the plus-strand) or negative-sense (called the minus-strand), depending on whether it is complementary to the viral messenger RNA (mRNA). Positive-sense viral RNA is identical to viral mRNA and thus can be immediately translated by the host cell. Negative-sense viral RNA is complementary to mRNA and thus must be converted to positive-sense RNA by an RNA polymerase before translation. DNA nomenclature is similar to RNA nomenclature, in that the *coding strand* for the viral mRNA is complementary to it ("), and the *non-coding strand* is a copy of it (+).

Genome size varies greatly between species. The smallest viral genomes code for only four proteins and have a mass of about 10^6 Daltons; the largest have a mass of about 10^8 Daltons and code for over one hundred proteins. RNA viruses generally have smaller genome sizes than DNA viruses due to a higher error-rate when replicating, and have a maximum upper size limit. Beyond this limit, errors in the genome when replicating render the virus useless or uncompetitive.

To compensate for this, RNA viruses often have segmented genomes where the genome is split into smaller molecules, thus reducing the chance of error. In contrast, DNA viruses generally have larger genomes due to the high fidelity of their replication enzymes.

Viruses undergo genetic change by several mechanisms. These include a process called genetic drift where individual bases in the DNA or RNA mutate to other bases. Most of these point mutations are "silent"—they do not change the protein that the gene encodes—but others can confer evolutionary advantages such as resistance to antiviral drugs. Antigenic shift is where there is a major change in the genome of the virus. This occurs as a result of recombination or reassortment.

When this happens with influenza viruses, pandemics may result. RNA viruses often exist as quasispecies or swarms of viruses of the same species but with slightly different genome nucleoside sequences. Such quasispecies are a prime target for natural selection. Segmented genomes confer evolutionary advantages; different strains of a virus with a segmented genome can shuffle and combine genes and produce progeny viruses or (offspring) that have unique characteristics.

This is called reassortment or *viral sex*. Genetic recombination is the process by which a strand of DNA is broken and then joined to the end of a different DNA molecule. This can occur when viruses infect cells simultaneously and studies of viral evolution have shown that recombination has been rampant in the species studied. Recombination is common to both RNA and DNA viruses.

Replication Cycle

Viral populations do not grow through cell division, because they are acellular; instead, they use the machinery and metabolism of a host cell to produce multiple copies of themselves, and they *assemble* in the cell.

A Typical Virus Replication Cycle

The life cycle of viruses differs greatly between species but there are six *basic* stages in the life cycle of viruses:

- *Attachment* is a specific binding between viral capsid proteins and specific receptors on the host cellular surface. This specificity determines the host range of a virus. For example, HIV infects only human T cells, because its surface protein, gp120, can interact with CD4 and receptors on the T cell's surface. This mechanism has evolved to favour those viruses that only infect cells in which they are capable of replication. Attachment to the receptor can induce the viral-envelope protein to undergo changes that results in the fusion of viral and cellular membranes.

- *Penetration* follows attachment; viruses enter the host cell through receptor mediated endocytosis or membrane fusion. This is often called viral entry. The infection of plant cells is different to that of animal cells. Plants have a rigid cell wall made of cellulose and viruses can only get inside the cells following trauma to the cell wall. Viruses such as tobacco mosaic virus can also move directly in plants, from cell-to-cell, through pores called plasmodesmata. Bacteria, like plants, have strong cell walls that a virus must breach to infect the cell. Some viruses have evolved mechanisms that inject their genome into the bacterial cell while the viral capsid remains outside.
- *Uncoating* is a process in which the viral capsid is degraded by viral enzymes or host enzymes thus releasing the viral genomic nucleic acid.
- *Replication* involves synthesis of viral messenger RNA (mRNA) for viruses except positive sense RNA viruses, viral protein synthesis and assembly of viral proteins and viral genome replication.
- Following the *assembly* of the virus particles, post-translational modification of the viral proteins often occurs. In viruses such as HIV, this modification, (sometimes called maturation), occurs *after* the virus has been released from the host cell.
- Viruses are *released* from the host cell by lysis—a process that kills the cell by bursting its membrane. Enveloped viruses (*e.g.*, HIV) typically are released from the host cell by budding. During this process the virus acquires its envelope, which is a modified piece of the host's plasma membrane.

The genetic material within viruses, and the method by which the material is replicated, vary between different types of viruses.

DNA Viruses

The genome replication of most DNA viruses takes place in the cell's nucleus. If the cell has the appropriate receptor on its surface, these viruses enter the cell by fusion with the cell membrane or by endocytosis. Most DNA viruses are entirely dependent on the host cell's DNA and RNA synthesising machinery, and RNA processing machinery.

The viral genome must cross the cell's nuclear membrane to access this machinery.

RNA Viruses

These viruses are unique because their genetic information is encoded in RNA. Replication usually takes place in the cytoplasm. RNA viruses can be placed into about four different groups depending on their modes of replication. The polarity (whether or not it can be used directly to make proteins) of the RNA largely determines the replicative mechanism, and whether the genetic material is single-stranded or double-stranded. RNA viruses use their own RNA replicase enzymes to create copies of their genomes.

Reverse Transcribing Viruses

These replicate using reverse transcription, which is the formation of DNA from an RNA template. Reverse transcribing viruses containing RNA genomes use a DNA intermediate to replicate, whereas those containing DNA genomes use an RNA intermediate during genome replication. Both types use the reverse transcriptase enzyme to carry out the nucleic acid conversion. Retroviruses often integrate the DNA produced by reverse transcription into the host genome.

They are susceptible to antiviral drugs that inhibit the reverse transcriptase enzyme, *e.g.* zidovudine and lamivudine. An example of the first type is HIV, which is a retrovirus. Examples of the second type are the *Hepadnaviridae*, which includes Hepatitis B virus.

Effects on the Host Cell

The range of structural and biochemical effects that viruses have on the hosts cell is extensive. These are called *cytopathic effects*. Most virus infections eventually result in the death of the host cell.

The causes of death include cell lysis, alterations to the cell's surface membrane and apoptosis. Often cell death is caused by cessation of its normal activities due to suppression by virus-specific proteins, not all of which are components of the virus particle.

Some viruses cause no apparent changes to the infected cell. Cells in which the virus is latent and inactive show few signs of infection and often function normally. This causes persistent infections and the virus is often dormant for many months or years. This is often the case with herpes viruses. Viruses, such as Epstein-Barr virus often cause cells to proliferate without causing malignancy, but viruses, such as papillomaviruses are an established cause of cancer.

Classification

Classification seeks to describe the diversity of viruses by naming and grouping them based on similarities. In 1962, Andre Lwoff, Robert Horne, and Paul Tournier were the first to develop a means of virus classification, based on the Linnaean hierarchical system. This system bases classification on phylum, class, order, family, genus, and species. Viruses were grouped according to their shared properties (not of their hosts) and the type of nucleic acid forming their genomes. Later the International Committee on Taxonomy of Viruses was formed.

ICTV Classification

The International Committee on Taxonomy of Viruses (ICTV) developed the current classification system and wrote guidelines that put a greater weight on certain virus properties to maintain family uniformity. A universal system for classifying viruses, and a unified taxonomy, has been established since 1966.

The 7th lCTV Report formalised for the first time the concept of the virus species as the lowest taxon (group) in a branching hierarchy of viral taxa. However, at present only a small part of the total diversity of viruses has been studied, with analyses of samples from humans finding that about 20% of the virus sequences recovered have not been seen before, and samples from the environment, such as from seawater and ocean sediments, finding that the large majority of sequences are completely novel.

The general taxonomic structure is as follows:

- Order (-virales)
- Family (-viridae)
- Subfamily (-virinae)
- Genus (*-virus*)
- Species (*-virus*).

In the current (2008) ICTV taxonomy, five orders have been established, the Caudovirales, Herpesvirales, Mononegavirales, Nidovirales, and Picornavirales.

The committee does not formally distinguish between subspecies, strains, and isolates. In total there are 5 orders, 82 families, 11 subfamilies, 307 genera, 2,083 species and about 3,000 types yet unclassified.

Baltimore Classification

The Nobel Prize-winning biologist David Baltimore devised the Baltimore classification system. The ICTV classification system is used in conjunction with the Baltimore classification system in modern virus classification.

The Baltimore classification of viruses is based on the mechanism of mRNA production. Viruses must generate mRNAs from their genomes to produce proteins and replicate themselves, but different mechanisms are used to achieve this in each virus family. Viral genomes may be single-stranded (ss) or double-stranded (ds), RNA or DNA, and may or may not use reverse transcriptase (RT). Additionally, ssRNA viruses may be either sense (+) or antisense (-). This classification places viruses into seven groups:

- dsDNA viruses (*e.g.* Adenoviruses, Herpesviruses, Poxviruses)
- ssDNA viruses (+)sense DNA (*e.g.* Parvoviruses)
- dsRNA viruses (*e.g.* Reoviruses)

- (+)ssRNA viruses (+)sense RNA (*e.g.* Picornaviruses, Togaviruses)
- (-)ssRNA viruses (-)sense RNA (*e.g.* Orthomyxoviruses, Rhabdoviruses)
- ssRNA-RT viruses (+)sense RNA with DNA intermediate in life-cycle (*e.g.* Retroviruses)
- dsDNA-RT viruses (*e.g.* Hepadnaviruses).

As an example of viral classification, the chicken pox virus, varicella zoster (VZV), belongs to the order Herpesvirales, family *Herpesviridae*, subfamily *Alphaherpesvirinae*, and genus *Varicellovirus*. VZV is in Group I of the Baltimore Classification because it is a dsDNA virus that does not use reverse transcriptase.

AGRICULTURAL IMPACT OF TRANSGENIC PLANTS

Outcrossing of transgenic plants not only poses potential environmental risks, it may also trouble farmers and food producers. Many countries have different legislations for transgenic and conventional plants as well as the derived food and feed, and consumers demand the freedom of choice to buy GM-derived or conventional products.

Therefore, farmers and producers must separate both production chains. This requires coexistence measures on the field level as well as traceability measures throughout the whole food and feed processing chain. Research projects such as Co-Extra, SIGMEA and Transcontainer investigate how farmers can avoid outcrossing and mixing of transgenic and non-transgenic crops, and how processors can ensure and verify the separation of both production chains.

BIOTECHNOLOGY AIDS PLANT BREEDING

Biotechnology is already becoming a powerful presence in our lives, yet most of us know very little about this important scientific phenomenon. Some of us are somewhat put off by technology, and sometimes frightened by it, but we can't keep ignoring it. The purpose of this article is to explain in simple terms what biotechnology is and to illustrate some of the positive aspects of biotechnology that are already benefitting, or are expected to benefit, the New Zealand fruitgrower.

How does this technology work? All living things carry their genetic inheritance in chemically coded messages or DNA (deoxyribonucleic acid). Under the microscope, DNA looks like a mass of tangled threads. But these threads consist of tiny subunits of DNA called genes. Genes carry instructions, sometimes called "the blueprint of life", for things such as fruit size or skin colour. Genes programme these instructions through arrangements of just four simple molecules-adenine, cytosine, thymine, and guanine. These A, C, T, and G's are like letters of the alphabet-their order "spells" out the language of life.

In nature genes (*i.e.* pieces of DNA) sometimes "mutate", *i.e.*, alter their chemical structure spontaneously from time to time. These mutations may result in plant cultivars carrying valuable new traits. For hundreds of years, farmers have taken advantage of natural mutations to improve traditional plant lines. For example, the popular apple cultivar "Royal Gala" is a natural mutation of "Gala". However, the time taken from a chance natural mutation to the production of a valuable new cultivar can be many years. But with genetic engineering, scientists are now able to shorten this time considerably.

The term "genetic engineering", sometimes also called "genetic manipulation" or "recombinant DNA", describes how scientists insert new genes into a plant and change its genetic makeup. This process is something like tearing out a page of the instruction manual for one living thing and glueing it into the manual of another.

The resulting plant which contains a new gene is called "transgenic". In the past, we could breed using only closely related species and it took many generations to introduce significant improvements. Genetic engineering allows researchers to shorten this time considerably by inserting only the gene responsible for a desired trait (*e.g.*, pest or disease resistance, skin colour, or fruit size) into a plant, and selecting for that trait.

This mimics the evolutionary process but in a highly selective way. Biotechnology here refers to the commercial application of genetic engineering to the development of new plants. Biotechnology could play a key role in improving our food supply.

It could provide us with more nutritious crops and longer-lasting food products that are easier to store and handle. Many genetically engineered plants are already growing in test fields overseas: canola oil with a healthier fat profile; a tomato with longer shelf life and better flavour; alfalfa and cotton with resistance to powerful herbicides; and cotton with resistance to insect pests.

SOIL MICROBES

These are soil fungi of which mushrooms and toadstools are the visible forms. And also the single-celled soil bacteria. You need a good compound microscope with high magnification to see bacteria and the single filaments of soil fungi.

SOIL MICROBES OVERVIEW

Soil looks lifeless but a quarter of a teaspoon of soil (1 gram) contains many microbes - 100 million bacteria, 1 million actinomycetes and 5 metres of fungal filaments.

Soil is certainly living! In pastures grazed by sheep near Armidale, NSW, Australia, we have measured the weight of microbes in soil and calculated that there is the equivalent weight of soil microbes of 88 sheep/ha.

above ground

below ground

Since the pastures were grazed with only 20 sheep/ha above-ground, there was the equivalent weight of 4 sheep (as microbes) in the soil to every real sheep grazing above.

Scientists in England and New Zealand have calculated that they have equivalent weights of 110 sheep/ha in microbes in pasture soil. These English and New Zealand soils would be moister and have more organic matter food for microbes than our Australian pasture soil which has periodic droughts and where organic soils are rare.

Ways of Feeding

Bacteria and fungi have similar ways of feeding. Their main food is organic detritus in leaf litter, soil, dung and carrion. In contrast to invertebrate animals which engulf food and digest it within their bodies, microbes digest their food outside themselves by secreting digestive enzymes over their food. Under the action of the enzymes, organic matter breaks down into simpler molecules external to the microbe, which then re-absorbs the simpler products of digestion through its cell wall.

Microbes have a vast number of enzymes to do this job – 50 to 60 different ones have been discovered. Few invertebrates possess this many digestive enzymes.

FUNGI

Most of the time fungi hide away in the murky depths of the soil and are not visible. But their microscopic, mouldy strands are growing through bits of organic matter and soil, through leaf litter, dead wood and dung and into carcasses of dead animals. Here, the fungi quietly go about their job of decomposing detritus and recycling the nutrients back to the soil.

It is only when these microscopic strands merge together to form the fruiting bodies of mushrooms and toadstools that we become aware of this invisible network beneath our feet. The spores borne on these fruiting bodies disperse in the wind, or in or on, the bodies of animals feeding on them e.g. springtails, flies.

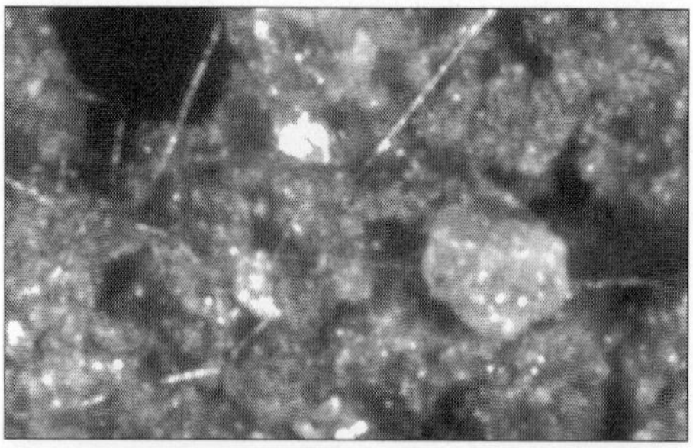

Fig. Microscopic Fungal Strands Growing Through Soil.

Fungal fruiting bodies are of many different and beautiful forms: mushrooms, toadstools, bracket fungi, puffballs, earthstars, coral fungi, large boletes and delicate crinoline fungi. Fungal fruiting bodies can also be rather gross. The starfish fungus smells revolting so as to attract blowflies which feed on the squishy mess in the centre of the fungus and transport its spores from place to place. Other soil fungi don't produce such conspicuous fruiting bodies, but form their spores or fruiting bodies within the soil e.g. the edible and delicious truffle which has to be sniffed out with the truffle-hunter's pigs or dogs.

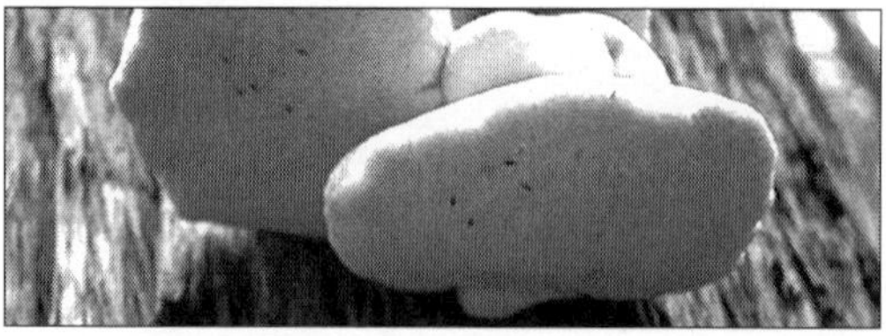

Fungi grow by dividing and spreading out into a network of branches which grow through organic matter. Fungi can link different parts of the soil profile together.

They can grow out from a piece of leaf litter, down into the soil, up into a dung pat and link all these together. It can absorb nutrients in one spot and transport them to another. This process is called translocation of nutrients and it is important in nutrient cycling. The networks of fungal filaments (mycelia) can explore their environment much more effectively than can soil bacteria which occur either as individual cells or as sedentary colonies.

The main food of fungi is organic matter in leaf litter, dung, soil and carrion. They (along with bacteria) are the main decomposer groups in soil because of their vast collective mass supported by the variety of digestive enzymes they can secrete onto organic matter to break it down.

Some fungi are edible and collecting wild fungi is a favourite pastime in Europe where, over the centuries, they have sorted out, by trial and error, those that are safe to eat - presumably the "errors" died! Australian fungi are a bit more problematical as we have not built up such a long-standing bed of knowledge. There are no simple tests for edibility.

Some old wives tales are: they are harmless if you can peel them, if they fail to turn dark on soaking in cold water or if a silver spoon cooked with the fungi, doesn't turn black. I wouldn't depend on these guidelines for edibility. To put these rules of thumb on a more scientific basis, don't eat mushrooms with yellow or greenish coloured gills, or have caps with yellow centres or caps which are thimble-shaped when young, but which smell of disinfectant (iodine) when stored in a plastic bag. Even then, it is best to stick to the fungi you can buy in the supermarket - there are many varieties to choose from these days.

Some soil fungi form a positive relationship with plant roots. They live in roots, but also extend their filament networks out into the soil where they absorb nutrients and water which they bring back to the plant to use. What the fungus gets out of this relationship is food in the form of sugars. This mutually beneficial relationship (or symbiosis) between root and fungus is called a mychorrizal association

The biggest organism in the world is invisible!

Not all soil fungi are goodies. The potato famine in Ireland in the mid-1800's was caused by a soil fungus. One in every nine people perished from starvation. Also, some soil-borne decomposer fungi can grow in grain and peanuts when these are stressed by drought. The fungus contaminates the grain and nuts with a toxin (aflatoxin) which causes liver damage in animals and humans.

Poisonous fungi

Only a small proportion of Australian fungi are poisonous but since some fungi have fatal effects it's best to stick to the supermarket varieties. Recently, a prominent Australian newspaper in its lifestyle section had to quickly publish a correction when it got the captions to two fungi mixed up. The paper had incorrectly labelled the lethal fly agaric as "edible"!

Quirky bit

Some microscopic fungi eat animals. Hard to imagine but the nematode-eating fungi grow little loops which soil nematode worms swim through. The loop tightens like a lassoo around the worm and the fungus oozes digestive enzymes onto it before absorbing the products of digestion into itself. Other nematode-eating fungi produce sticky knobs which stick to and trap nematodes.

BACTERIA

Bacteria are tiny one-celled micro-organisms, 0.001 mm in diameter, which need a light microscope with very high magnification to see them. They are best seen with an electron microscope.

Bacteria have a different way of growing to fungi. They multiply by cell division, one cell dividing to make 2 cells and so on. Sometimes the cells move apart and they live as separate individuals or they can stay together to form a colony (a few cubic mm in volume) as they do when cultured in the laboratory on an agar plate. Fungal cells are different. They divide and form branching networks of filaments (mycelia) in the soil and can explore the soil environment much more effectively than bacteria.

Bacteria form resistant spores when the climate turns harsh. They can survive droughts in a spore stage and when favourable conditions return, they can germinate and grow again.

The main food of bacteria is organic matter in leaf litter, dung, soil and carrion. They (along with fungi) are the main decomposer groups in soil because of the huge number of digestive enzymes they can secrete onto organic matter to break it down.

Some soil bacteria can fix atmospheric nitrogen, ultimately making it available to plants. Some of these live free in the soil e.g. Azotobacter, while others fix nitrogen while living in nodules on the roots of legumes e.g. Rhizobia.

Here, these bacteria gain food in the form of sugars from the plants and the plant benefits through increased nitrogen fixed by the bacteria. This "win-win" relationship between bacteria and plant is called "symbiosis" - or a relationship where both organisms gain a mutual benefit from living together.

Diseases from soil bacteria

Not all soil bacteria are "goodies" as soil-borne diseases can attack plants and animals. Anthrax and tetanus, which infect domestic animals and humans, lurk in soil as does the bacterial blight of tomatoes.

OTHER MICROBES

Other microbes, apart from bacteria and fungi, also live in soil. These are actinomycetes, yeasts, viruses and algae.

Not all of these other microbes are decomposer organisms:

- Actinomycetes are bacteria-like microbes, joined end to end in a filamentous strand to form a branched, fungus-like network. They seem to be half-way between a bacterium and a fungus. They are good decomposers and can break down particularly resistant forms of organic matter such as cellulose and chitin.
- Soil yeasts are unicellular fungi. They are decomposers.
- Soil viruses prey on soil bacteria but not a lot is known about these micro-organisms.
- Soil algae are also included in the microbes but these are different as they photosynthesise like green plants do.

Quirky bit

Oddly enough, soil actinomycetes are a good source of antibiotics. The soil actinomycete, Streptomyces, has produced many of our antibiotics and drugs e.g. streptomycin. The antiparasitic drug, avermectin, was first discovered in an actinomycete living in soil on a golf course in Japan! It is used to protect domestic animals, pets and humans from both internal and external parasites such as roundworms, ticks and lice. In humans, it kills parasitic nematode worms which cause River Blindness in Africa and South America.

BENEFITS OF MICROBES

Significant resources have been spent to understand and fight disease-causing microorganisms from the beginning of microbiology. You may be surprised to learn that only a small faction of microbes are involved in disease; many other microbes actually enhance our well-being.

The harmless microbes that live in our intestines and on our skin actually help us fight off disease. They actively antagonize other bacteria and take up space, preventing potential pathogens from gaining a foothold on our bodies.

Indeed, like all large organisms, humans have entire communities of microorganisms in their digestive systems that contribute to their overall health. The microbial community in humans not only protects us from disease, but also provides needed vitamins, such as B_{12}. Human health and nutrition also depends on healthy farm animals. Cows, sheep, horses and other ruminant animals utilize their microbial associates to degrade plant material into useful nutrients.

Without these bacteria inside ruminants, growth on plant material would be impossible. In contrast to humans ruminant animals have complex stomachs that harbour large numbers of microorganisms. These microbes degrade the plant stuff eaten by the animal into usable nutrients. Without the assistance of the microbes, ruminant animals would not be able to digest the food they eat. Commercial crops are also central to human prosperity, and much of agriculture depends upon the activities of microbes. For example, an entire group of plants, the legumes, forms a cooperative relationship with certain bacteria.

These bacteria convert nitrogen gas to ammonia for the plant, an important nutrient that is often limiting in the environment. Microbes also serve as small factories, producing valuable products such as cheese, yogurt, beer, wine, organic acids and many other items. In conclusion, while it is less apparent to us, the positive role of microbes in human health is at least as important as the negative impact of pathogens. A picture of nodulated leguminous plants. In this case pea plants. Nodules are visible on the roots of the plants in the left of the picture. The plants on the right were not inoculated with nodulating bacteria.

ENVIRONMENT EFFECTED BY MICROBES

The vast majority of life on this planet is microscopic. These teaming multitudes profoundly influence the make-up and character of the environment in which we live. Presently, we know very little about the microbes that live in the world around us because less than 2 per cent of them can be grown in the laboratory. Understanding which microbes are in each ecological niche and what they are doing there is critical for our understanding of the world. Microbes are the major actors in the synthesis and degradation of all sorts of important molecules in environments.

Cyanobacteria and algae in the oceans are responsible for the majority of photosynthesis on Earth. They are the ultimate source of food for most ocean creatures (including whales) and replenish the world's oxygen supply. Cyanobacteria also use carbon dioxide to synthesize all of their biological molecules and thus remove it from the atmosphere.

Since carbon dioxide is a major greenhouse gas, its removal by cyanobacteria affects the global carbon dioxide balance and may be an important mitigating factor in global warming. In all habitats, microorganisms make nutrients available for the future growth of other living things by degrading dead organisms. Microbes are also essential in treating the large volume of sewage and wastewater produced by metropolitan areas, recycling it into clean water that can be safely discharged into the environment.

Less helpfully (from the view of most humans), termites contain microorganisms in their guts that assist in the digestion of wood, allowing the termites to extract nutrients from what would otherwise be indigestible. Understanding of these systems helps us to manage them responsibly and as we learn more we will become ever more effective stewards. Energy is essential for our industrial society and microbes are important players in its production.

A significant portion of natural gas comes from the past action of methanogens (methane-producing bacteria). Numerous bacteria are also capable of rapidly degrading oil in the presence of air and special precautions have to be taken during the drilling, transport and storage of oil to minimize their impact. In the future, microbes may find utility in the direct production of energy. For example, many landfills and sewage treatment plants capture the methane produced by methanogens to power turbines that produce electricity.

Excess grain, crop waste and animal waste can be used as nutrients for microbes that ferment this biomass into methanol or ethanol. These biofuels are presently added to gasoline and thus decreasing pollution. They may one-day power fuel cells in our cars, causing little pollution and having water as their only emission.

Finally, We are increasingly taking advantage of the versatile appetite of bacteria to clean up environments that we have contaminated with crude oil, polycholrinated biphenyls (PCBs) and many other industrial wastes. This process is termed bioremediation and is a cheap and increasingly effective way of cleaning up pollution.

MICROBES AROUND US

Microorganisms will grow on simple, cheap medium and will often rise to large populations in a matter of 24 hours. It is easy to isolate their genomic material, manipulate it in the test tube and then place it back into the microbe. Due to their large populations it is possible to identify rare events and then, with the use of powerful selective techniques, isolate interesting bacterial cells and study them.

These advantages have made it possible to test hypothesis rapidly. Using microbes scientists have expanded our knowledge about life. Below are a few examples. Microorganisms have been indispensable instruments for unlocking the secrets of life. The molecular basis of heredity and how this is expressed as proteins was described through work on microorganisms. Due to the similarity of life at the molecular level, this understanding has helped us to learn about all organisms, including ourselves. Some prokaryotes are capable of growing under unimaginably harsh conditions and define the extreme limits of where life can exist. Some species have been found growing at near 100 °C in hot springs and well above that temperature near deep-sea ocean vents.

Others make their living at near 0 °C in freshwater lakes that are buried under the ice of Antarctica. The ability of microbes to live under such extreme conditions is forcing scientists to rethink the requirements necessary to support life. Many now believe it is entirely possible that Jupiter's moon Europa may harbour living communities in waters deep below its icy crust. What may the rest of the universe hold? Until recently, while we could study specific types of bacteria, we lacked a cohesive classification system, so that we could not readily predict the properties of one species based on the known properties of others. Visual appearance, which is the basis for classification of large organisms, simply does not work with many microbes because there are few distinguishing characteristics for comparison between species even under the microscope. However, analysis of their genetic material in the past 20 years has allowed such classification and spawned a revolution in our thinking about the evolution of bacteria and all other species. The emergence of a new system organizing life on Earth into three domains is attributable to this pioneering work with microorganisms. The fruits of basic research on microbes has been used by scientists to understand microbial activity and therefore to shape our modern world.

Human proteins, especially hormones like insulin and human growth factor, are now produced in bacteria using genetic engineering. Our understanding of the immune system was developed using microbes as tools. Microorganisms also play a role in treating disease and keeping people healthy. Many of the drugs available to treat infectious disease originate from bacteria and fungi. One last recent role of microbes in informing us about our world has been the tools they provide for molecular biology. Enzymes purified from bacterial strains are useful as tools to perform many types of analyses. Such analyses allow us to determine the complete genome sequence of almost any organism and manipulate that DNA in useful ways.

We now know the entire sequence of the human genome, with the exception of regions of repetitive DNA, and this will hopefully lead to medical practices and treatments that improve health. We also know the entire genome sequence of many important pathogens. Analysis of this data will eventually lead to an understanding of the function of critical enzymes in these microbes and the development of tailor-made drugs to stop them. The tools of molecular biology will also affect agriculture.

For example, we now know the complete genome sequence of the plant *Arabidopsis* (a close relative of broccoli and cauliflower). This opens a new avenue to a better understanding of all plants and hopefully improvements in important crops. Microbes have a profound impact on every facet of human life and everything around us. Pathogens harm us, yet other microbes protect us. Some microbes are pivotal in the growth of food crops, but others can kill the plants or spoil the produce.

Bacteria and fungi eliminate the wastes produced in the environment, but also degrade things we would rather preserve. Clearly they effect many things we find important as humans. In the remainder of this chapter we take a look at how scientists came to be interested in microbes and follow a few important developments in the history of microbiology.

2

The Epidemiology of Foodborne Disease

INTRODUCTION

Recent outbreaks of foodborne disease caused by many different pathogens and involving a variety of food products have been the subject of headlines, but it is unclear whether the incidence of foodborne disease has increased over the last generation. The major reason for the uncertainty is that the lack of a national foodborne-disease surveillance system has prevented the study of trends in disease rates. What is clear is that factors affecting the potential safety of the nation's food supply have changed dramatically over the last generation and justify concern that the incidence of foodborne disease is high and may be increasing.

At least five trends contribute to the possible increase in foodborne disease: changes in diet, the increasing use of commercial food services, new methods of producing and distributing food, new or re-emerging infectious foodborne agents, and the growing number of people at high risk for severe or fatal foodborne diseases.

Diet

Annual food expenditures in the United States, as a share of disposable personal income, decreased from 14 percent to 11 percent from 1970 to 1996. No industrialized nation spends a smaller share of its wealth on food than the United States. For much of the population, readily available food is more varied and more affordable than ever before. For example, in the 1960s, an average US grocery store had fewer than 7,000 food items available. Today, an average US grocery store sells about 30,000 food items, and over 12,000 new products are introduced each year.

During this time when relative costs of the US food supply are decreasing, per capita consumption of many foods has changed substantially. Public health efforts to promote a "heart-healthy" diet have helped to boost the consumption of fresh fruits and vegetables. On a per capita basis, in 1995, Americans ate about 31 lb more commercially grown vegetables, including potatoes and sweet potatoes, and 24 lb more fresh fruit than in 1970. As the consumption and variety of produce have increased, so has the importation of produce from developing countries. The General Accounting Office estimates that in 1995 one-third of all fresh fruit consumed in the United States was imported. Food imports have increased both because of lower production costs in foreign countries and because of consumer demand for year-round supplies of fruits and vegetables that have limited growing seasons in the United States.

For example, 17 percent of cantaloupes, 52 percent of green onions, 36 percent of cucumbers, and 34 percent of tomatoes sold in the United States in 1996 were grown in Mexico. Seasonally, as much as 79 percent of a particular commodity consumed in this country has been raised in Mexico alone, and the percentage of produce from other developing countries consumed here is growing rapidly. Fresh produce items were the leading vehicle associated with foodborne disease outbreaks in Minnesota from 1990 to 1996, accounting for almost one-third of all outbreaks. This percentage is higher than that available from national foodborne disease surveillance data and possibly reflects more active surveillance in that state.

Other trends in the United States are the decreasing consumption of beef and the increased consumption of chicken and seafood. In 1970, the average American consumed 79 lb of beef, 27 lb of chicken, and 12 lb of fish and shellfish; in 1996, annual per capita consumptions were 64, 50 and 15 lb, respectively. Contamination of red meat with *Salmonella and Escherichia coli* 0157:H7 remains important, but the risk posed by chicken as a vehicle for *Campylobacter and Salmonella* has grown substantially.

Cultural changes affect not only what Americans eat, but also where they eat and how their food is prepared. Increasingly, Americans have time-pressured lifestyles. Saving time and effort in shopping for and preparing food will continue to be important for many Americans. Households with a single parent or two working adults often face particular pressures in food shopping and meal preparation that can affect food selection and safety. Cookbook sales and television cooking shows demonstrate that cooking is popular, but there seems to be decreased interest in ordinary daily home food preparation and, with most adults working, a lack of role models in the kitchen. Reductions in time for and interest in home food preparation also result in changed food patterns, fewer homemade dishes, more reliance on leftovers, increased purchase of prepared or convenience foods, and frequent eating away from home.

With less time spent in the kitchen and greater availability of high-quality, ready-to-eat dishes and convenience items, Americans' food preparation skills are diminishing. As a result, appreciation of simple but critical food safety techniques, such as washing hands and utensils and storing foods at optimal temperatures, has likely diminished.

Individual food tastes and preparation styles are brought to the United States from around the world, and the increasing ethnic diversity of the

American population may affect food safety in several ways. Different ethnic groups have different concerns and practices regarding food safety, and this could affect the activities of immigrants as food preparers at home and in the workplace. New food risks can arise as immigrant populations adapt traditional preferences and practices to their new environment. Food safety education programmes should consider the food beliefs and practices of various cultural groups.

Although data show a rise in perceived risk of foodborne illness among consumers, attitudes do not always translate into improved food-handling practices.

Over half of all shoppers report washing their hands and/or food preparation surfaces, yet only 28 percent know that cooking temperatures are critical and that foods should be refrigerated promptly. A recent Food and Drug Administration study found that many US consumers still eat undercooked hamburger meat and raw eggs.

Commercial Food Services

The percentage of the food dollar spent on food consumed away from home has risen dramatically over the last three decades. In 1970, only 34 percent of our food dollars were spent eating away from home. In contrast, in 1996, 46 percent of our food dollars were spent for meals and snacks prepared outside the home.

Institutional feeding sites serve a wide range of people—from very young children in child care centres to the elderly—in congregate sites for meals, alternative-care centres, and nursing homes. Meals are also prepared as takeout foods from supermarkets and convenience stores; many of these meals include one or more cold food items, such as delicatessen sandwiches and salads that require extensive food handling and are not cooked before consumption. These changes have led to an increase in the number of people handling food and the potential for an increase in the transmission of foodborne diseases from food handlers to consumers.

The average food handler in this country earns the minimum wage, lacks sick leave and other health benefits, and has very limited opportunities for advancement.

These jobs are filled by people with few employment opportunities and low economic status, conditions that may be related to a high incidence of intestinal diseases and low rates of routine hand washing. Recent evidence from Minnesota demonstrates the increased risk of food handler associated transmission of *Salmonella typhimurium;* this finding was documented because *Salmonella* isolates in Minnesota undergo molecular characterization and epidemiologic investigation. Food handlers in other areas of the United States probably play a similar role in the transmission of *Salmonella* and other enteric agents.

Production and Distribution

The changing availability and sources of our food supply have brought changes in methods of producing and distributing food. Today, large manufacturing plants can process quantities and types of products that two decades ago would have required many smaller plants. Although the ability to control possible hazards increases when there are fewer plants, the potential for larger outbreaks—even if product contamination is minimal—is evident. Thousands of people become ill during such outbreaks. For example, contamination of tanker trucks with nonpasteurized liquid eggs and subsequent use of those trucks to haul pasteurized ice cream mix most likely led to the largest documented outbreak of salmonellosis in the United States. That outbreak showed that even sporadic low level contamination of a single product can result in a major epidemic of foodborne illness because of the quantity of product consumed.

In the spring of 1998, a cereal produced by the largest generic label manufacturer of cereal in the United States caused a national outbreak of *Salmonella agona* infection. Outbreaks such as this can be difficult to detect because contamination of the product is sporadic and the product is marketed widely.

The driving forces of globalization, advanced technology, and economic competitiveness have dramatically affected the structure and practices of livestock and poultry production. Herds, flocks, and other populations of food animals, including fish and seafood, are increasingly concentrated in fewer and larger production units. The traditional farmstead model of the past, often characterized by multiple species and small numbers of food animals reared on a single farm, has been replaced by specialized, large-scale production systems. For example, from 1994 to 1995, the number of US hog operations decreased from 207,980 to 182,700, but both the inventory of hogs and the total number of hogs produced increased during the same time. That mirrors the general agribusiness trend, and the restructuring continues.

Infectious Foodborne Agents

The role of new or re-emerging causes of foodborne diseases is well-recognized, but the size of the increase in risk is unknown. Emerging and reemerging infections have been defined as new, recurring, or drug-resistant infections whose incidence in humans has increased in the last two decades or whose incidence threatens to increase in the near future.

Recent examples of newly recognized agents are *E. coli* 0157:H7, other pathogenic *E. coli, Cyclospora, and Cryptosporidium.* Old agents re-emerging in new vehicles or product streams include *Salmonella enteritidis* in eggs and hepatitis A virus in produce. Finally, there have been recent increases in antimicrobial resistance of pathogens, such as *S. typhimurium* DT104 and *Campylobacter jejuni* in humans, which have been attributed to the use of antimicrobial agents in food animals. The frequently cited annual estimates of foodborne disease (up to 81 million cases) and 9,000 associated deaths are based on assumptions that do not necessarily reflect the current national foodborne disease problem.

Those estimates must be qualified for two reasons. First, no comprehensive population-based studies of gastrointestinal illness in the community have attempted to determine what proportion of these illnesses is due to consumption of contaminated food and what proportion is from other sources. Second, foodborne illness can cause clinical conditions not characterized by gastrointestinal symptoms, such as congenital toxoplasmosis, hemolytic uremic syndrome, salmonella-associated septicemia, and invasive *Listeria* infections. In the absence of comprehensive estimates of foodborne disease incidence, the only data available are those from the FoodNet programme.

Data from FoodNet-an effort sponsored by the Centres for Disease Control and Prevention, the US Department of Agriculture, FDA, and the health departments of California, Connecticut, Georgia, Minnesota, and Oregon-have demonstrated that the incidence of diarrheal illness in the United States is about 1.4 episodes per person per year, or some 370 million episodes of diarrhea each year.

Population-based studies of diarrheal disease conducted from 1948 through 1971 in Ohio and Michigan demonstrate an incidence of about one episode per person per year. The reason for the apparent 40 percent difference between these estimates is unclear, but it is evident that the problem is large. If only 25 percent of diarrhoeal disease is food-related, the burden of foodborne diarrheal disease in the United States far exceeds current estimates.

Studies conducted by FoodNet suggest that fewer than 2 percent of the cases of diarrheal illness in the United States can be attributed to *Salmonella, Campylobacter, Shigella, E. coli* 0157:H7, and *Yersinia*. Little information is available on the role of other pathogens, including newly recognized agents such as Shiga toxin-producing *E. coli* other than 0157:H7, *Cryptosporidium, and Cyclospora,* and previously established pathogens such as Norwalk-like viruses, *Clostridium perfringens, Staphylococcus aureus, Bacillus cereus,* and hepatitis A virus. Until additional studies elucidate the role of the recognized pathogens in foodborne disease, we can only speculate about the occurrence of new or unrecognized pathogens.

FATAL FOODBORNE DISEASE

The risk of foodborne disease is related to several factors, including the presence and dose of a pathogen or toxin in food, the virulence of the pathogen or toxin, the mechanisms of transmission, and the susceptibility of a host. Many factors influence susceptibility to infection and the severity of disease, including age, the use of immunosuppressive agents, and disease states that increase immuno-suppression.

Young children are more likely than adults to develop illness from selected pathogens, and children have been high-profile victims of several foodborne disease outbreaks in recent years. Young children today are likely to be exposed to a broader range of foodborne diseases than was true a generation ago, because families eat out or take prepared food home more often and children receive more community and out-of-home care at younger ages.

Elderly persons are particularly susceptible to illness from foodborne disease, and the number of the elderly in the United States is increasing rapidly. From 1965 to 1995, the number of Americans aged 65 years or older grew by 82 percent. As the 75 million persons in the "baby boomer" generation (equaling one-third of the population) age, the United States can expect one fifth of the population to be over the age of 65 within the next three decades.

As the population ages, the incidence of cancer and other chronic diseases is likely to increase. These diseases will require an increased use of chemotherapeutic regimens, drugs to deal with rejection of solid organ transplantations, and antimicrobial drugs that have important effects on the normal (and beneficial) bacterial flora of the intestinal tract. Together, these factors predispose people to both the occurrence and the serious outcomes of foodborne disease. The increasing number of people with immunosuppressive diseases, such as human immunodeficiency virus, also contributes to the public health importance of food safety.

CHEMICAL HAZARDS ASSOCIATED WITH THE FOOD SUPPLY

Changes in our society and our food supply have raised new concerns about food chemical safety. Some of the changes that have raised new concerns about foodborne infectious disease are also affecting how government agencies carry out the more familiar task of protecting the food supply from toxic chemical agents.

Foods are themselves a complex collection of naturally occurring chemicals that have nutritive, organoleptic, and pharmacological functions and occasionally toxic effects. Naturally occurring toxicants probably present risks second only to those imposed by microorganisms. These natural toxicants include seafood toxins and foodborne mycotoxins.

Other chemicals are introduced into foods intentionally or unintentionally. Intentionally added substances include food and colour additives, flavours, enzymes, vitamins and minerals, and other ingredients that help to add value or characteristic properties or functions to a food. Because of broad public concern about synthetic chemicals, the toxicologic profiles of many of these synthetic components of food are much more complete than those of natural components of food. As a result, the risks posed by regulated food additives are generally better characterized than those of many naturally occurring substances.

Unintentional additives can include environmental or industrial contaminants as well as some substances used in food production but not intended to be part of the food. The migration of food production substances into the final product is generally very low, but must be carefully regulated to ensure safety.

Examples include sanitizers used to keep food production surfaces safe, packaging materials used to keep food safe and fresh, pesticides used on crops and drugs used in animals to mitigate damage, disease, microbial toxin production, and general food losses. Environmental or industrial contaminants are not sanctioned but have the potential to enter the food chain. Examples of chemical contamination incidents are methyl mercury in fish, mistaken mixing of polybrominated biphenyls (a fire retardant) into animal feed, and leakage of ammonia refrigerant into frozen foods.

Some people are sensitive or allergic to chemical constituents that are harmless to the rest of the population. These allergic reactions are estimated to occur in 1 to 2 percent of adults and 5 to 8 percent of children. Although serious reactions are rare, it is estimated that several dozen deaths occur each year because of allergic reactions to food.

More than 160 foods and food-related substances have been identified as being able to cause allergic reactions. However, in the United States, more than 90 percent of allergic reactions appear to be caused by just eight food types: peanuts, tree nuts (such as walnuts, pecans, almonds, hazel nuts), crustacea (shrimp, lobster, and other shellfish), eggs, milk, soy, fish, and wheat.

Consequently, the subject of chemical hazards in food is complex and includes consideration of potential risks that vary widely in scope and severity. Many chemical hazards associated with foods have been recognized only in the last century as advancements have been made in chemistry, toxicology, and risk assessment.

Public concern over these hazards has grown in recent decades, in line with the increasing distrust of chemicals generally and of their use in the environment. Indeed, there have been episodes of chemical intoxication (usually arising out of accident or occupational exposure) with tragic consequences. Although most experts agree that the more serious hazards in the American food supply are not chemical but microbiological, public concern has demanded that proportionally more regulatory resources be applied to chemical hazards.

New Food Components

The American public's growing interest in the relationship between diet and health has led to an increased demand for foods or food constituents that not only have nutritive value but also hold promise for prevention or even treatment of disease. These products have been referred to as dietary supplements, functional foods, pharma-foods or nutraceuticals. In a recent study, more than two-thirds of surveyed households reported use of a vitamin, mineral, or herbal supplement within the previous six-month period. The Dietary Supplement Health and Education Act of 1994 eased restrictions on certain statements of nutrition support made for supplements and exempted them from the safety approval requirement applicable to conventional food additives.

That legal change helped spur the growing market for supplements of all types, including herbal products, and raises food safety concerns. Some supplements and herbal products on the market may pose a risk of adverse health effects because they are not required to meet specified safety standards before being sold. They may thus contain varying amounts or unknown or inadequately characterized ingredients that can have pharmacological activity that has not been adequately characterized.

Food processors are examining relatively new sources of ingredients for more conventional functional properties. For example, gums and fibers such as konjac flour, tara gum, inulin, or psyllium fiber may provide bulk and texture to foods, yet have had limited food use in the United States. Potential broad use of these ingredients in domestic foods raises important safety questions: are there significant intestinal effects such as blockage or reduced transit times? Are there impacts on vitamin uptake from the intestinal tract or on osmotic balance? What are the potential allergy risks from increased exposure to plant-sourced materials? These are a few of the safety questions that may be raised when a new ingredient source is considered.

Food processors are also utilizing macronutrient substitutes, such as nonnutritive sweeteners and fat replacers in many food products. These macronutrient substitutes have potential value by lowering calories, sugar, or fat in food products. Because dietary quantities of these substitutes could be substantially higher than the amounts of typical food additives, the assessment of their safety can be particularly important as well as particularly difficult.

New Food Technologies

Modification of plants or animals via genetic engineering can improve yields and increase resistance to pests. This new technology might offer improvements in food safety through increased resistance to molds that produce food mycotoxins or through lower levels of allergenic proteins, fatty acids, or other undesirable components of food. However, there are important differences between countries in how food products from genetically modified organisms are regulated.

Several products derived from genetic engineering have been declared safe by US regulators, but many European countries have either forbidden their sale or insisted on what marketers believe is disparaging labeling. Concerns about the safety of products from genetically engineered plants and animals are only partially resolved. Food irradiation is not a new technology.

Irradiation of fruits and vegetables decreases the risk of pathogens and extends shelf life. But as discussed, public concerns about the safety of irradiated food, fueled in part by a lack of consumer education, limit its use in the United States. These concerns prompted the requirement that certain irradiated foods be labeled as such. The extent to which this labeling requirement has limited subsequent adoption of the process is unknown.

Re-emerging Toxic Agents

As the science of toxicology has progressed over the last 50 years, questions about the safety of chemicals in foods have become more sophisticated. When tougher safety standards are applied to new food chemicals or reapplied to old ones, new issues of toxicity emerge. The cycle of reevaluating safety standards began in the years after World War II when advances in the understanding of the mechanisms of carcinogenesis coincided with the increased use of rodent bioassays. The newly focused attention to carcinogens that ensued led Congress to pass the 1958 "Delaney clause" proscription of carcinogens.

The focus on carcinogens in food continues to consume substantial testing and regulatory resources. Further advances in biomedical research have raised new concerns and new standards of safety that now address teratogenicity, reproductive toxicity, mutagenicity, hormonal effects, and immunotoxicity.

The effects of food constituents on hormonal function is of concern as medical research has indicated the importance of certain hormones in regulating diseases such as breast cancer and osteoporosis. This phenomenon, sometimes referred to as endocrine disruption, led Congress in 1996 to mandate special regulatory attention to the issue and inspired the Environmental Protection Agency (EPA) to convene an expert panel, the Endocrine Disrupter Screening and Testing Advisory Committee. That committee has attempted to define endocrine disrupters, determine analytical methods, and recommend how the disrupters should be regulated. Potential candidates for evaluation include constituents of some food packaging materials, pesticides, and natural food constituents, such as genestein in soy.

Some groups have recently asked whether chemical safety assessments provide adequate safety margins to protect children. That concern draws attention because new pesticide legislation directs EPA to consider the aggregate risks posed by any pest control agent; that is, exposures should be assessed for all potential sources of a chemical and for other chemicals with a similar mechanism of action. Another concern is the emergence of evidence of subtle developmental and behavioural effects of relatively low concentrations of chemicals (for example, lead).

In summary, chemical hazards, as well as foodborne pathogens, present new and changing challenges to the food safety system. Federal food safety efforts will need full integration and sufficient support to meet these challenges.

Physical Hazards

The foods we consume begin as raw agricultural commodities grown in open fields or waters or raised in a variety of production facilities, such as barns, coops, pens, and feedlots. Rocks, stones, metal, wood, glass, and other physical objects can become part of raw ingredients. Further contamination can occur in the transport, processing, or distribution of foods because of equipment failure, accident, or negligence.

Foreign physical materials in foods can cause serious harm to consumers. Protective devices that remove or prevent physical hazards include metal detectors, magnets, sieves, traps, scalpers, and screens. Other effective means of protection are production plant policies against the use of glass, wood, or nonferrous metal where possible; employee training; quality audits of ingredient suppliers; and sensory tests. Federal agencies have established "defect action levels" for natural or unavoidable defects that do not affect human health, such as stems or pits in fruits and vegetables, bone in mechanically deboned meats, and microscopic insect fragments, which are primarily of aesthetic concern and do not present a safety hazard.

FOODBORNE PATHOGENS

Foodborne pathogens are the leading causes of illness and death in less developed countries, killing approximately 1.8 million people annually. In developed countries, foodborne pathogens are responsible for millions of cases of infectious gastrointestinal diseases each year, costing billions of dollars in medical care and lost productivity. New foodborne pathogens and foodborne diseases are likely to emerge, driven by factors such as pathogen evolution, changes in agricultural and food manufacturing practices, and changes to the human host status. There are growing concerns that terrorists could use pathogens to contaminate food and water supplies in attempts to incapacitate thousands of people and disrupt economic growth.

Enteric Viruses

Food and waterborne viruses contribute to a substantial number of illnesses throughout the world. Among those most commonly known are hepatitis A virus, rotavirus, astrovirus, enteric adenovirus, hepatitis E virus, and the human caliciviruses consisting of the noroviruses and the Sapporo viruses. This diverse group is transmitted by the fecal-oral route, often by ingestion of contaminated food and water.

Protozoan Parasites

Protozoan parasites associated with food and water can cause illness in humans. Although parasites are more commonly found in developing countries, developed countries have also experienced several foodborne outbreaks. Contaminants may be inadvertently introduced to the foods by inadequate handling practices, either on the farm or during processing of foods. Protozoan parasites can be found worldwide, either infecting wild animals or in water and contaminating crops grown for human consumption. The disease can be much more severe and prolonged in immuno-compromissed individuals.

Mycotoxins

Molds produce mycotoxins, which are secondary metabolites that can cause acute or chronic diseases in humans when ingested from contaminated foods. Potential diseases include cancers and tumors in different organs (heart, liver, kidney, nerves), gastrointestinal disturbances, alteration of the immune system, and reproductive problems.

Species of Aspergillus, Fusarium, Penicillium, and Claviceps grow in agricultural commodities or foods and produce the mycotoxins such as aflatoxins, deoxynivalenol, ochratoxin A, fumonisins, ergot alkaloids, T-2 toxin, and zearalenone and other minor mycotoxins such as cyclopiazonic acid and patulin. Mycotoxins occur mainly in cereal grains (barley, maize, rye, wheat), coffee, dairy products, fruits, nuts and spices.

Control of mycotoxins in foods has focused on minimizing mycotoxin production in the field, during storage or destruction once produced. Monitoring foods for mycotoxins is important to manage strategies such as regulations and guidelines, which are used by 77 countries, and for developing exposure assessments essential for accurate risk characterization. Aflatoxins are still recognized as the most important mycotoxins.

They are synthesized by only a few Aspergillus species, of which A. flavus and A. parasiticus are the most problematic. The expression of aflatoxin-related diseases is influenced by factors such as age, nutrition, sex, species and the possibility of concurrent exposure to other toxins. The main target organ in mammals is the liver, so aflatoxicosis is primarily a hepatic disease. Conditions increasing the likelihood of aflatoxicosis in humans include limited availability of food, environmental conditions that favour mold growth on foodstuffs, and lack of regulatory systems for aflatoxin monitoring and control.

Yersinia Enterocolitica

Yersinia enterocolitica includes pathogens and environmental strains that are ubiquitous in terrestrial and fresh water ecosystems. Evidence from large outbreaks of yersiniosis and from epidemiological studies of sporadic cases has shown that Y. enterocolitica is a foodborne pathogen. Pork is often implicated as the source of infection. The pig is the only animal consumed by man that regularly harbors pathogenic Y. enterocolitica.

An important property of the bacterium is its ability to multiply at temperatures near 0°C, and therefore in many chilled foods. The pathogenic serovars (mainly O:3, O:5, 27, O:8 and O:9) show different geographical distribution. However, the appearance of strains of serovars O:3 and O:9 in Europe, Japan in the 1970s, and in North America by the end of the 1980s, is an example of a global pandemic. There is a possible risk of reactive arthritis following infection with Y. enterocolitica.

Vibrio

Vibrio species are prevalent in estuarine and marine environments, and seven species can cause foodborne infections associated with seafood. Vibrio cholerae O1 and O139 serovtypes produce cholera toxin and are agents of cholera. However, fecal-oral route infections in the terrestrial environment are responsible for epidemic cholera. V. cholerae non-O1/O139 strains may cause gastroenteritis through production of known toxins or unknown mechanism.

Vibrio parahaemolytitcus strains capable of producing thermostable direct hemolysin (TDH) and/or TDH-related hemolysin are most important causes of gastroenteritis associated with seafood consumption. Vibrio vulnificus is responsible for seafoodborne primary septicemia, and its infectivity depends primarily on the risk factors of the host.

V. vulnificus infection has the highest case fatality rate (50%) of any foodborne pathogen. Four other species (V. mimicus, V. hollisae, V. fluvialis, and V. furnissii) can cause gastroenteritis. Some strains of these species produce known toxins, but the pathogenic mechanism is largely not understood. The ecology of and detection and control methods for all seafoodborne Vibrio pathogens are essentially similar.

Staphylococcus Aureus

Staphylococcus aureus is a common cause of bacterial foodborne disease worldwide. Symptoms include vomiting and diarrhea that occur shortly after ingestion of S. aureus toxin-contaminated food. The symptoms arise from ingestion of preformed enterotoxin, which accounts for the short incubation time. Staphylococcal enterotoxins are superantigens and, as such, have adverse effects on the immune system.

The enterotoxin genes are accessory genetic elements in S. aureus, meaning not all strains of this organism are enterotoxin-producing. The enterotoxin genes are found on prophages, plasmids, and pathogenicity islands in different strains of S. aureus. Expression of the enterotoxin genes is often under the control of global virulence gene regulatory systems.

Campylobacter

Campylobacter spp., primarily C. jejuni subsp. jejuni is one of the major causes of bacterial gastroenteritis in the U.S. and worldwide. Campylobacter infection is primarily a foodborne illness, usually without complications; however, serious sequelae, such as Guillain-Barre Syndrome, occur in a small subset of infected patients. Detection of C. jejuni in clinical samples is readily accomplished by culture and nonculture methods.

Listeria Monocytogenes

Listeria monocytogenes is Gram-positive foodborne bacterial pathogen and the causative agent of human listeriosis. Listeria infections are acquired primarily through the consumption of contaminated foods, including soft cheese, raw milk, deli salads, and ready-to-eat foods such as luncheon meats and frankfurters.

Although L. monocytogenes infection is usually limited to individuals that are immunocompromised, the high mortality rate associated with human listeriosis makes it the leading cause of death among foodborne bacterial pathogens. As a result, tremendous effort has been made to develop methods for the isolation, detection and control of L. monocytogenes in foods.

Salmonella

Salmonella serotypes continue to be a prominent threat to food safety worldwide. Infections are commonly acquired by animal to human transmission though consumption of undercooked food products derived from livestock or domestic fowl. The second half of the 20th century saw the emergence of Salmonella serotypes that became associated with new food sources (*i.e.* chicken eggs) and the emergence of Salmonella serotypes with resistance against multiple antibiotics.

Shigella

Shigella species are members of the family Enterobacteriaceae and are Gram negative, nonmotile rods. Four subgroups exist based on O-antigen structure and biochemical properties: S. dysenteriae (subgroup A), S. flexneri (subgroup B), S. boydii (subgroup C) and S. sonnei (subgroup D). Symptoms include mild to severe diarrhea with or without blood, fever, tenesmus and abdominal pain.

Further complications of the disease may be seizures, toxic megacolon, reactive arthritis and hemolytic uremic syndrome. Transmission of the pathogen is by the fecal-oral route, commonly through food and water.

The infectious dose ranges from 10-100 organisms. Shigella spp. have a sophisticated pathogenic mechanism to invade colonic epithelial cells of the host, man and higher primates, and the ability to multiply intracellularly and spread from cell to adjacent cell via actin polymerization. Shigella spp. are one of the leading causes of bacterial foodborne illnesses and can spread quickly within a population.

Escherichia Coli

More information is available concerning Escherichia coli than any other organism, thus making E. coli the most thoroughly studied species in the microbial world. For many years, E. coli was considered a commensal of human and animal intestinal tracts with low virulence potential. It is now known that many strains of E. coli act as pathogens, inducing serious gastrointestinal diseases and even death in humans.

There are six major categories of E. coli strains that cause enteric diseases in humans, including the:

- Enterohemorrhagic E. coli, which cause hemorrhagic colitis and hemolytic uremic syndrome,
- Enterotoxigenic E. coli, which induce traveller's diarrhea,
- Enteropathogenic E. coli, which cause a persistent diarrhea in children living in developing countries,
- Enteroaggregative E. coli, which provokes diarrhea in children,
- Enteroinvasive E. coli that are biochemically and genetically related to Shigella species and can induce diarrhea,
- Diffusely adherent E. coli, which cause diarrhea and are distinguished by a characteristic type of adherence to mammalian cells.

Clostridium Botulinum and Clostridium Perfringens

Clostridium botulinum produces extremely potent neurotoxins that result in the severe neuroparalytic disease, botulism. The enterotoxin produced by C. perfringens during sporulation of vegetative cells in the host intestine results in debilitating acute diarrhea and abdominal pain. Sales of refrigerated, processed foods of extended durability including sous-vide foods, chilled ready-to-eat meals, and cook-chill foods have increased over recent years. Anaerobic spore-formers have been identified as the primary microbiological concerns in these foods.

Heightened awareness over intentional food source tampering with botulinum neurotoxin has arisen with respect to genes encoding the toxins that are capable of transfer to nontoxigenic clostridia.

Bacillus Cereus

The Bacillus cereus group comprises six members: B. anthracis, B. cereus, B. mycoides, B. pseudomycoides, B. thuringiensis and B. weihenstephanensis. These species are closely related and should be placed within one species, except for B. anthracis that possesses specific large virulence plasmids. B. cereus is a normal soil inhabitant, and is frequently isolated from a variety of foods, including vegetables, dairy products and meat. It causes a vomiting or diarrhea illness that is becoming increasingly important in the industrialized world. Some patients may experience both types of illness simultaneously.

The diarrheal type of illness is most prevalent in the western hemisphere, whereas the emetic type is most prevalent in Japan. Desserts, meat dishes, and dairy products are the foods most frequently associated with diarrheal illness, whereas rice and pasta are the most common vehicles of emetic illness.

The emetic toxin (cereulide) has been isolated and characterized; it is a small ring peptide synthesised nonribosomally by a peptide synthetase. Three types of B. cereus enterotoxins involved in foodborne outbreaks have been identified. Two of these enterotoxins are three-component proteins and are related, while the last is a one-component protein (CytK). Deaths have been recorded both by strains that produce the emetic toxin and by a strain producing only CytK.

Some strains of the B. cereus group are able to grow at refrigeration temperatures. These variants raise concern about the safety of cooked, refrigerated foods with an extended shelf life. B. cereus spores adhere to many surfaces and survive normal washing and disinfection (except for hypochlorite and UVC) procedures. B. cereus food borne illness is likely under-reported because of its relatively mild symptoms, which are of short duration.

FOOD AUTHENTICITY

It is important to be able to detect microorganisms in food, in particular pathogenic microorganisms or genetically modified microorganisms. Real-time PCR is an accepted analytical tool within the food industry. Its principal role has been one of assisting the legislative authorities, major manufacturers and retailers to confirm the authenticity of foods. The most obvious role is the detection of genetically modified organisms, but real-time PCR makes a significant contribution to other areas of the food industry, including food safety.

EFFECTS OF NUTRITION LABELING ON FOOD FORMULATION

While discussions about the Nutrition Facts box typically revolve around its impact as a tool to help consumers make more healthful food selections, it must be recognized that the regulations governing the Nutrition Facts box and the associated nutrient content claims also influence the formulation of products. Manufacturers often adjust the quantities of particular ingredients or discretionary fortificants so that their products can be shown in the Nutrition Facts box to have a higher percent DV for some nutrients and a lower percent DV for others, thereby meeting the criteria for particular content claims.

Thus any changes to the DV or to the list of nutrients included in the Nutrition Facts box can be expected to have some effect on the nutrient profiles of processed food. Furthermore, implementation of the recommended principles for discretionary fortification is expected to affect the inclusion of nutrients and their amounts suitable for fortification.

Overages

In the United States, for the purpose of determining compliance with nutrition labeling regulations, nutrients added to fortified or fabricated food. A food containing a Class I nutrient is deemed to be misbranded if the amount of the nutrient in a composite sample (collected and analyzed in accordance with regulations) is not at least equal to the value declared on the label. This requirement differs from that for Class II nutrients, which are those that naturally occur (i.e., are indigenous) in food. The nutrient content of a composite sample containing a Class II nutrient must be equal to at least 80 percent of the value declared on the label.

In order to ensure compliance with label declarations, fortified nutrients are often added in excess (an overage). The amount of overage to ensure compliance depends on several factors, including the chemical stability of the nutrient itself, the manufacturing process (e.g., where in the process a vitamin or mineral is added; how well the vitamin or mineral is incorporated into the product; the conditions of time, temperature, pressure, and moisture), and the conditions used to simulate abusive handling throughout the distribution and retail chain (because manufacturers cannot control conditions after a product leaves their factories and distribution centres). In the United States reasonable excesses of vitamins and minerals over labeled amounts are acceptable within current good manufacturing practices.

In attempting to comply with the regulation for Class I nutrients, some manufacturing practices may result in unnecessary, excessive overages. Excessive overages would be of concern for those nutrients with a low margin between the DV and the lowest UL and for which a serious adverse effect is the basis for the UL. Even in the absence of the potential for an adverse effect, excessive overages, which may not be captured in food composition databases, complicate the evaluations of nutrient intakes and nutritional status.

Public Health Benefit in Nutrition Labeling

The tone of the message conveyed by the elements in the Nutrition Facts box merits careful consideration because the box serves as an important public health communication tool. When the Nutrition Facts box is revised, the committee suggests that thought be given to the selection, organization, and display of nutrients as these elements may impact the tone of the public health message.

The Nutrition Facts box currently can be construed as presenting a negative message because many of the required nutrients that appear in bold print on the top of the Nutrition Facts box (e.g., cholesterol, fat, and sodium) are those that consumers are expected to restrict in order to reduce their risk of chronic disease.

There is no similar emphasis made by grouping, format, or letter size of those nutrients for which consumers are encouraged to increase their intake (e.g., calcium). The priorities of required nutrient selection, label design, and other factors need to be reviewed in light of the potential positive message tone and educational value that could be presented for nutrients included on the label.

In 1973 the selection of nutrients and food components to be included on nutrition labeling was primarily based on ameliorating nutritional deficiencies and on illustrating the positive and negative nutrient content of food. In 1990 FDA critically reviewed these nutrients, modified the list, and placed more emphasis on food components associated with chronic diseases and less emphasis on nutrient-deficiency diseases.

In particular the revision placed emphasis on those nutrients that reflected the primary public health objective of a reduction in the risk of cardiovascular disease and the secondary objective of a reduction in the risk of cancer.

THE RELATIONSHIP BETWEEN FOOD AND HEALTH

It is also clear that since the 1970s important changes have occurred in popular understanding of the relationship between food and health. Preventative medicine may only be in its infancy, and receive little official encouragement, but for some groups of people in the industrialized world, healthy eating is now considered essential.

Healthy eating in the past depended critically on local custom and diet and, most importantly, income. Today this is still true, but local variations have less importance, and additional factors have made their appearance, which have engaged the attention of a battery of professionals, such as health educators, doctors, consumer groups and alternative therapists. The perceived need to diet to reduce obesity has given rise to a huge consumer market in low-calorie foods, including the controversial 'low-cal' liquid diets and innumerable diet 'systems'.

The publication of evidence about heart disease and cancer has led to some marked shifts in the public consumption of certain foods, especially in the United States, where the changes date back to the publication of the McGovern Report in 1977. Food has become one of the obsessions of our time, associated as it is with good health, longevity and the occurrence of stress. Physical fitness has also become an obsession in some quarters, perhaps also a 'fad'. Evidence from the United States suggests that, while in 1960 only a quarter of adults exercised regularly, by 1980 over half did so.

The point is not that everybody is exercising more, but that more people are, and that health is increasingly linked to diet in many people's minds. This is not a trend likely to be discouraged by the food industry, since it also represents a change with commercial possibilities. Indeed, the giant food firms have jumped on this bandwagon, adapting to each new nutritional recommendation—low sodium, low saturated fats, high fibre—to the point of confusing consumers grappling with labels and elusive media-speak, such as 'lite' and 'natural'.

Changes have also occurred in public perceptions of the two processes: appropriation and substitution. Following the scare over DDT between 1969 and 1972 in the United States, and subsequent publicity about the use of other pesticides, attention has been given to the effects of pesticide residues in food, especially by the London Food Commission in the United Kingdom. Few studies have been undertaken, and even fewer given publicity, about the relative nutritional quality of organically produced and chemically produced food. Pesticide use in the United States rose 500 per cent between 1950 and 1986, but in the latter year one-fifth of United States's crops were lost to pests, the *same percentage as in* 1950! The tactic employed by the food industry in meeting the criticism of environmentalists and whole food enthusiasts, has been that the industry can 'compensate' through industrial processes for the shortcomings of nature.

The industry argued that organic foods were more expensive (ignoring the hidden subsidies provided to chemically produced food), that organic foods could not meet market demand, that food additives increased the palatability of food, and that convenience removed the drudgery from food preparation. By the late 1960s in North America, and perhaps a decade later in the United Kingdom, food manufacturers began to 'reposition' themselves, as well as their food products.

Now they were on the side of emancipated women and environmentalists; they offered themselves as accomplices in women's drive for more independence and in the vanguard of responsible, sustainable resource management. Market researchers have credited much of the improvement in the food industry to their own responsiveness to consumer pressure. In the words of an industry spokesperson:

For example, ten years ago the only place people could find additive-free *natural* products were in the comparatively cramped and small premises of health food stores. Today such products, not only much improved and—dare we say it?—engineered to satisfy consumer preferences, are mainstream products in prime placing in major supermarket chains the length and breadth of Britain.

This represents one side of the story, but in practice healthier eating has also presented a challenge to the food industry which it has found difficult to grasp. The problem, in a nutshell, is how to leave out food additives and processing, and still add value to the final product.

The post-war period had seen additives 'substitute' for natural ingredients, while much that was natural had been removed from food. Food was given more value added through packaging, processing and ensuring it stayed 'fresh' longer.

The challenge represented by making 'healthy' foods from 'health' foods was how to make profits by appearing to do less to the product. The industry responded to the challenge, as we have noted, by both market and product differentiation—identifying, and exploiting market opportunities for 'niche' products, such as low-calorie foods, ethnic health foods and fitness-related foods.

Finally, the food industry has been in the forefront of the promotion of food supplements; people can now supplement their inadequate diet from a range of commercially promoted food 'accessories'. In 1985 over US$3 billion was spent in the United States on vitamin and mineral supplements and over $4.5 billion in supplements to fortified breakfast cereals.

This diet supplementation was the inevitable outcome of the way eating had changed in the United States. In 1909, 40 per cent of calories had been provided from fruit, vegetables and grains. By 1976 only 20 per cent of calories came from these sources, the rest from fats and refined sugars. As people ate more calories they ate fewer vitamins and minerals—supplements have helped to fill the vacuum left by changes in the 'affluent diet' of North America.

Another important social change that has accompanied the shifts in diet referred to above is that governments in the industrialized countries have been forced to make some accommodation to the pressures put upon them by food consumers, and interest groups representing consumers. In the United Kingdom this has come about largely as a result of successive food 'scares', such as those over salmonella, listeria and bovine spongiform encephalopathy ('mad cow disease') in the last few years. The extent to which governments will respond to consumer anxieties is still unclear, but the current debate about public nutrition policy and food legislation represents an excellent illustration of the kind of contradiction to emerge from the development of the modern food system.

The demand for more intervention in the food system presents difficulties for governments committed to the idea of choice in the marketplace. On the one hand it is difficult for them to resist, as a matter of principle, the demand for more information about food products. Issuing 'guidelines' about food and health is not inconsistent with conservative ideology, which places individual choice before government controls. The publicity given to the McGovern Report in the United States after 1977 served to divide the food lobby. Part of the food industry could take comfort from the encouragement being given to people to eat more cereals, vegetables and fruit. Another section of the industry tried to develop marketing strategies to overcome opposition to increased consumption of other items (particularly meat, dairy products and eggs).

The development of a 'fordist' diet in the 1950s and 1960s had meant that red meat consumption rose, in the United States from 121 Ibs per capita in 1930 to 160 Ibs per capita in 1970. By 1968 20 million tons of vegetable protein was being fed to livestock to produce 2 million tons of animal protein. Since 1980 there has been a steady reversal of this trend: beef consumption has fallen in the United States, and voluntary reductions in food consumption, like those recommended by the McGovern Commission for eggs, sugar and animal fats, have also begun to be registered. However, this, too, is only part of the total picture.

Changes in diet, encouraged by public debate, have not yet pushed or persuaded governments, in the United States or Western Europe, to take a more interventionist stand on other aspects of food consumption, notably the use of food additives, although recent legislation like that proposed in the United Kingdom in March 1990 would introduce more rigorous standards of hygiene in the food industry. In the United States the Federal Food and Drug Administration (FDA) co-operates very closely with the food industry. In practice many of its recommendations are welcomed by the industry because they reinforce the idea that most food already on the supermarket shelves is fit for public consumption.

The FDA-initiated testing of food led to the conclusion that over a thousand substances used as food additives were 'generally recognized as safe' (GRAS) for use in food, whereas only two substances were banned as a result of FDA testing: a carcinogenic weed killer and, in 1969, cyclamates. The United States Government has proved reluctant to act as a watchdog on food additives. Meanwhile over 1 billion Ibs of additives are added to food products in the United States, equivalent to 5 Ibs of additives per capita.

ASPECT OF FOOD CONSUMPTION

It is important to consider another aspect of food consumption: the meaning people attach to it, and the way in which food consumption practices help structure social relations more generally. Gofton has made a useful first attempt to come to grips with some aspects of the sociology of food consumption in the United Kingdom today.

He notes several of the trends referred to above and in other chapters of this book: the growing concern for health; home-centred living; redefinitions of social roles within households; the trade-off between the time taken in food preparation and leisure activities; and the emphasis given in our society to individualism and hedonism.

Many of these changes are linked to the way society has become organized in the developed countries, such as the shift towards a more informational society where the production of information rather than goods is at a premium. Similarly, there has also been a shift towards relatively decentralized authority systems, particularly in the family, where individual members are more likely to make autonomous decisions about (among other things) what and when they eat. Finally, during the 1980s we witnessed in the United Kingdom what has been dubbed the beginnings of 'the enterprise culture', which placed importance, particularly in ideological terms, on individual entrepreneurship and initiative rather than bureaucratic surveillance and government responsibility.

Gofton argues, convincingly we believe, that patterns of food consumption (and, we should add, patterns of resistance to highly processed, 'industrialized' foods) are closely related to changes in social behaviour and organization, which place private consumer choice and individualism before community-based or class-based patterns of behaviour. The trends in consumption, implied by different market 'niches', together with underlying shifts in the way households and 'communities' are constituted around specific interests, are important components of post-industrial society.

Gofton also reviews various strands of what we might term the 'sociology of food' focusing on three major areas in which attention to food might be linked to 'post-industrial' and 'post-modernist' concerns: the semiotics of food (represented by Mary Douglas in particular); the debate initiated by Pahl and Wallace and Gershuny about changes in the way the household is constituted; and the work of Bourdieu on the relationship between upbringing, class and taste (including taste in food). Something of this approach is reflected, within British anthropology, in the work of Jack Goody.

The work of Pahl and Gershuny, important as it is, is principally concerned with work rather than food, but the other sociological currents certainly reflect serious attention to food consumption habits, and the role of these habits in our culture. These approaches explore the phenomenology of food consumption, the way in which food tastes and habits operate not simply as indicators of lifestyle but also as ways of conveying meaning to experience. Other contributors to this debate include Charles and Kerr, whose recent work has addressed the relationship between eating habits, gender relations and the class structure, basing their analysis on empirical evidence drawn from interviews with over 200 women in the United Kingdom between 1982 and 1983.

Charles and Kerr take as their point of departure the way in which food practices contribute to the reproduction of the social order, drawing on Bourdieu and Althusser in stressing the role of ideology in social reproduction. Their analysis places considerable emphasis on the meanings carried by food about the nature of the social occasion, and the way these meanings provide the ground rules for social structure, gender roles and class identities. This kind of research is phenomenological in outlook but essentially empirical, even occasionally empiricist, in design. It is grounded less in what Giddens has termed the 'knowledge ability of [the] agents' themselves and more in the unconscious behaviour of these agents, whose cultural patterns help to define them within their society.

Research such as that undertaken by Charles and Kerr delineates some important aspects of social behaviour in Britain today, but we would argue that it fails to relate behaviour within the family to wider structural changes in the role of food within society, and to the expansion of the food system. Reading Charles and Kerr one could be excused for thinking that working-class people never ate 'fast foods' (there is no entry on fast foods in the index) while 'convenience foods' are relegated to just three pages of text. Apparently the women interviewed used convenience foods, but not in the main meal they prepared—convenience foods were not 'proper meals'. Expending time and effort on the production of a meal in some way conferred goodness on it: perhaps a moral rather than a nutritional goodness but this distinction was not made by the women...A meal consisting of convenience foods...cannot constitute a proper meal, is by definition less healthy than a proper meal and involves much less preparation time...We may therefore see women's concern that the family should eat properly and healthily in terms of home cooked food provided at regular intervals as having been handed down through family tradition; it is a central part of family life.

This kind of analysis places evidence of continuity before evidence of change, and imbue the meanings people attach to their behaviour with more importance than the behaviour itself.

It is also instructive to note how the authors have renegotiated these meanings in the text. It is not easy to establish from their work whether women are making more use of convenience foods, although the suggestion is that they probably are, because the analysis is largely concerned to demonstrate that any resort to convenience foods needs to be located within traditional family behaviour.

Emphasizing that people look upon food as 'natural' should not blind us to the fact that in practice it frequently is highly processed. What people *feel about food* may be important, but so is the evidence about *what they eat*. One suspects that if the same women were talking to researchers less concerned to establish its meaning for their argument about gender and class relationships, it would have been much more difficult to discuss food preferences in terms of the provision of a 'proper meal'.

Instead of evidence for the reproduction of the existing social order, we would see evidence of change, in the social changes surrounding the way food is consumed as well as in the food itself. Phenomenological approaches to the meaning of food consumption would benefit from taking some cognizance of the processes through which food itself has become an industrially produced good.

This book has examined the effects of the modern food system in accelerating social changes, in both developed and developing societies. We have argued that the ideological aspects of the food system are not confined to those usually reported on by sociologists: the 'appropriateness' of food, and the meaning attached to food in everyday life, within the experience of working-class people. It is our contention that changes in the material basis of food production and consumption are linked to the transformation of 'nature', and that the establishment of new areas of commodity production influences the values attached to social relations.

Capitalism is an evolving system, whose values and practices do not lie embedded in the past, anymore than the behaviour of working-class people is lodged in static definitions of what is an appropriate, definition formed by their parents and grandparents within a quite different social milieu. Food has become fully commoditized; it is therefore to be expected that it's 'naturalness' would be emphasized by the food industry, in the process generating new symbolic meanings as a path to further accumulation.

As Massey has observed there needs to be a debate about the 'natural', which can draw on recent post-structuralist thinking. The 'denaturalization' of space, in the form of attention to 'location theory', was a preoccupation of some social scientists, particularly geographers in the 1960s. What is required now is similar attention to the way in which naturalization accompanies changes in technology and social relations.

Nature has been refashioned in several ways, which this book has drawn attention to within an integrated framework of analysis:

1. Within the labour process the employment of women has, through routinization, become naturalized as women's labour has shifted into and out of the home.
2. Technological changes in agri-food production, now enhanced by biotechnology and genetic engineering, have manipulated biological processes, appropriating some and substituting others, redefining and refashioning nature as a source of profit and capital accumulation. As the food we consume has become more processed it has been presented as more 'natural' by the food industry.
3. Finally, the development of modern agriculture has forced a separation between the 'countryside' as an area of consumption (amenity, recreation, aesthetics) and the production of food and fibre from the land. The drive to protect the countryside, or more generally to seek greater protection for conservation and wilderness areas in other industrial societies, is defended and advanced on the basis of conserving what is 'natural', and a true embodiment or representation of 'nature'.

It is clear from this analysis that as the modern food system becomes more integrated, and the management of contradictions, such as those presented by the environment, become more urgent, the more the food system departs—in all three senses referred to above—from being 'natural'. Naturalization is the price society pays for seeking to transform its image of itself. Or, as Bourdieu has expressed it, each system produces the naturalization of its own arbitrariness.

THE MODERN FOOD SYSTEM

We have argued that many of the contradictions posed by the modern food system can be handled by policy interventions, such as those to contain environmental costs and to address public concern over the quality of industrially produced food. The modern food system developed out of a combination of factors. These included: state intervention in agricultural policy designed to stimulate the adoption of new technology; guaranteed markets; the food processing industries' access to female labour, partially released from full-time employment in the home by (initially) wartime restructuring; and changes in family farming which served to legitimate the demands of more 'efficient', small-scale but highly capitalized producers. The management of the contradictions of the modern food system can be handled in the industrialized countries of the North, at a price, for several reasons that require close attention.

As we have seen, productivity increases in agriculture have been subsidized through industrial accumulation in the countries of the North. In the post-war period of intensive accumulation, the significance of cheap food arises not simply, as initially understood from its role, as a wage good; but also from its contribution to a widening of the market for consumer goods in general, which has accompanied the growth of labour productivity within the advanced capitalist societies.

The costs of this policy, of continuing to support agricultural modernization, have been passed on to the consumer and taxpayer in the European Community, the United States and Japan. It is increasingly clear that this agricultural policy model, even when only partially understood, is being called into question, notably by conservative critics as well as those with 'Greener' credentials. At the level of ideological legitimation, which has remained important throughout the post-war period in helping to define the relationship between farmers and the state, the resistance of public opinion to the priority attached to this model, at great environmental and social cost, has become an increasingly important part of the new political agenda of the 1990s.

Furthermore, the rise to importance of what we have termed 'environmental managerialism', in the industrialized countries, represents a significant policy opening, which could not have been anticipated a decade ago. The improved management of 'externalities' associated with the food system can reduce the full impact of the transformation of the countryside, through negotiated accommodations to environmental concerns. In the United Kingdom we have been conditioned to accept a managed countryside for a least a century; in other countries of the European Community the emphasis on preserving the cultural association between family farmers and the environment is equally strong politically.

Indeed this nexus of interests recently gained prominence as the philosophical foundation of the EuropeanCommi-ssion's proposed rural development policy, as outlined in *The Future of Rural Society*. In North America, the issues are gather different, in that the separation of wild lands (to which public access is considered important) and agriculture, is more pronounced. Nevertheless, here too the armoury of interventionist policies designed to manage the contradictions of development is receiving more attention: low input farming, wilderness retention, the establishment of recreational priorities for an affluent, urban population.

Finally, as we have seen, in the industrialized countries of the North there has been considerable resistance to the complete industrialization of the diet. For a substantial minority of the population—in the United States usually considered to be about a third of the market—'healthy' eating has become an alternative to the excesses of the 'fordist' diet. The food industry has accommodated, indeed sought to exploit, many of these changes, whose origin we see in the wider educational opportunities of the post-war generation, as well as greater private affluence. As we have seen, it is possible to increase value added in food products without necessarily losing the nutritional value of food, although this poses real problems for the large food conglomerates in increasingly competitive, oligopolistic markets.

Within some sectors of the durable consumer goods industries the development of 'healthy' convenience foods can also provide a stimulus to production, microwaves and toasted sandwich-makers replace or supplement other white goods, but without leading to the consumption of more unhealthy food. Changes in diet have also been prompted by wider changes in public attitudes: the increases in vegetarianism, the concern with animal welfare, the widespread interest in high-fibre, low-calorie foods as part of attempts to lose weight. These changes are significant, and are most unlikely to be reversed: they have helped to set in motion a series of effects on the supply side of the food system, which are reverberating and will continue to reverberate during the remaining years of this century and into the next.

Within the developed countries of the North, then, the modern food system poses problems for the economy and society which call for changes in policy, and the management of the environment, but which can be successfully addressed, albeit at a price in both financial and political terms. The implications of the food system for the South, we would contend, are altogether more radical. It is worth considering the differences carefully.

In the South the modern food system can only be partially adopted, because structural inequalities have prevented the industrial sector from acting as the 'motor' for the restructuring of agriculture around industrially produced food for mass consumption. In most developing countries the agricultural sector, and particularly staple production, has been appallingly neglected, not least in terms of the impact of agricultural intensification on the poor and the environment.

Where rising agricultural productivity is supported by state subsidies, or subsidies from the developed countries and their institutions, it has benefited some groups much more than others. The 'fordist' diet, exemplified by the penetration of the grains-livestock complex, has only reached a very narrow band of food consumers in even the so-called 'middle-income' countries of Latin America. In most of Africa and Asia, the fordist diet remains part of the mystique of Western cultural values, as distant from its audience as the movie star. Environmental managerialism, too, does not represent a workable strategy for most governments in the South, where it has neither tradition nor institutional foundation.

In the North, the improved management of the environment, usually at the behest of relatively affluent, urban groups, is endorsed with relatively little political dissent; the modest interventions to ensure a better management of the countryside, or conservation of wilderness, have overwhelming public support in Europe and North America. In the South, on the other hand, land degradation is intimately linked to poverty in rural areas. A vast, largely rural population of poor people, many of them *very* poor, looks to the environment as the basis of its precarious livelihood.

Environmental problems in developing countries can only be realistically understood as the outcome of underdevelopment and poverty. The 'environment' lacks ideological resonance; it is to be 'tamed' or 'conquered', and made to yield economic revenues and livelihoods. Environmental managerialism, without its ideological underpinnings which explain its strong cultural appeal in the North, would need to be radically modified to make any impact on the causes of environmental degradation in the South.

It can be argued, indeed, that attempts to manage the environment in developing countries, in as much as they are predicated on Northern experience, are not merely unworkable, they are also frequently prejudicial to the interests of the poor. In the Brazilian Amazon the disputed land and natural resources cannot be 'mediated' successfully by international agencies alone, whose principal concern is thought by many Brazilians (on both left and right of the political spectrum) to be the protection of the quality of life in the North, rather than the rights to livelihood of poor people in the South.

Many Brazilians say, and their views are at least partly convincing, that campaigns to protect the Amazon rain forest are the new face of 'neo-imperialism', prompted by fears of global warming (largely exaggerated in view of the much more serious effect of fossil fuel emissions) and the perceived need to preserve rare animal and plant species, rather than concern about poor frontier settlers or the pernicious social impacts of the Brazilian debt.

In many parts of the South, 'resistance' is to be understood primarily in terms of the right to livelihood, to citizenship and to the recognition of basic human rights. Struggles such as those surrounding land invasions, or the protection of indigenous people and rubber tappers in the forests, represent struggles for existence, rather than movements to secure improvements in the quality of life. The development of the modern food system is linked to these struggles, albeit sometimes in tenuous ways, but the management of the environment cannot resolve the contradictions implied by the development process. Indeed, the effects of recent structural adjustment policy, and attempts to manage the enormity of the debt burden in Latin America and Africa in particular, have intensified environmental problems, and make any 'solution' that does not include major changes in these policies largely irrelevant. The implications of the partial adoption of elements of the modern food system in the South can only be understood, and the resulting problems successfully addressed, by changes in structural policies encompassing the global economy. They cannot be resolved, as to a large extent they can in the North, where they occur.

Finally, as far as changes in the diet are concerned, the partial development of the modern food system in the countries of the South has served to replace (and in some cases destroy) existing, indigenous food systems. This has occurred without effective resistance from a broadly based mass movement of consumers, educated as to their rights, and able to exert pressure on government to accede to their demands. It is virtually impossible to envisage public campaigns over salmonella or food additives in Mexico or Indonesia, or, what is altogether more telling, effective mass movements to introduce new standards of food hygiene in shops or on street corners. Meanwhile, food irradiation—increasingly contested by the radical food lobby in the North—opens up market opportunities in the South.

It is clear, then, that what is at issue if we consider North-South relations, and examine the implications of the development of the modern food system within an international framework, is a set of concerns which stretch the parameters of our model, and our analysis. To speak of resistance and struggle in the South, is to speak of the survival of alternative systems which link the production of food directly to its transformation and consumption, systems which embody quite different patterns of social reproduction, systems which—in the form of ubiquitous 'externalities'—carry many of the environmental and social costs of our 'successful' development model. Our ability to grasp the fact that the economic and political hegemony of our values, and our economic policies, have served to undermine the integrity of other food systems, remains the major challenge we in the North face today.

For the spread of the modern food system marks the success of the North's cultural hegemony, as well as its economic supremacy. We need to address the sustainability of our own models, then, before we are in a position to hand them down to others.

3

The Environment of Sustainable Agriculture

INTRODUCTION

Agriculture is directly linked to very many facets of sustainable development, including poverty eradication, sustainable consumption and production, management of natural resources, energy, freshwater, health, education, trade and market access, as well as technology transfer and capacity building.

Agriculture is an integral part of the general development system, serving the system as a whole, and being served by it. If the effects of other sections of the development systems reduce sustainability, then sustainability of agriculture is also affected. A sustainable system should be resilient, and able to withstand shocks and failures of parts of its systems without the whole collapsing, and without small shocks leading to a spiral of unsustainability.

Agriculture centres on integrated use of natural resources such as soil, water, climate and biological diversity.

The integration of agriculture with other aspects of land management and ecosystem conservation is essential in order to promote both environmental sustainability and agricultural production:

- Natural resources have to be accessible to the poor, which in the farm sector means secure rights to land, water and genetic resources. For this, there is a need to develop public-private partnerships.
- Policies for land and water resource management, biodiversity protection, infrastructure investment, strategy on institutional market reforms, reduction of tariffs and phasing out of possible commodity control are integral to achieving sustainable agriculture. ?Sustainability should be seen in the context of different agro-climatic zones as well of as the country as a whole. Suitable technologies should be developed and indicators for sustainable agriculture should be identified for both.
- It is essential to provide funding for integrated rural development

plans, programmes and strategies, at national and regional levels, with particular emphasis on investment in economic and social infrastructure in rural areas, enterprise development, human resource development, and capacity building for local governance.

- An efficient credit policy regime with the required rural banking and credit system will play a major role in the future. The government should encourage investment in vital agriculture infrastructure, credit linkages, and use of new and appropriate techniques towards this end.
- More than 70 per cent of the country's agriculture is under small and marginal farmers with limited resources. A nationwide crop insurance scheme will provide such farmers the needed confidence to invest and gain from technological advances in agriculture.
- The focus on accelerated food grains production on a sustainable basis and free trade in grains, as well as on rural employment opportunities will lead to faster economic growth and give purchasing power to the people, which in turn would help increase household food security.
- Concerted efforts should be made at national, regional and local levels to pool, distil and evaluate traditional practices, knowledge and wisdom and to harness them for sustainable agricultural growth.
- It should be recognized that information is a critical input for agricultural development. It is as important as other key inputs including credit, seeds, nutrients and water. Information can be efficiently converted into economically rewarding opportunities.
- It is critical to recognize that the challenge to world agriculture is both technological (requiring the development of new, high productivity, environmentally sustainable production systems), and political (requiring policies that do not discriminate against rural areas in general, and agriculture in particular).

Equitable Distribution and Access

A positive right to life would imply that the State provides to each and every person, adequate food and other basic necessities, and that it ensures a healthy environment, so that people may live and grow in dignity. The most important challenge in the 1980s and 1990s was physical access to food.

In this millennium, the challenge is economic and ecological access to food:

- A transition from chemical and machinery-intensive technologies to ecological farming technologies is required towards providing sustained physical access to food.
- Environmental access involves on the one hand, attention to soil health care, water harvesting and management, conservation of forests and biodiversity, and on the other hand to sanitation, environmental hygiene, primary health care and primary education.

- Emphasis on economic access underlines the need for promoting sustainable livelihoods through multiple income-earning opportunities.
- It is important to increase food availability in areas where it is produced, thus reducing transport costs and excessive dependence on international markets.
- Shift from existing expensive, inefficient and ineffective institutional arrangements, to decentralized management systems of food storage and distribution will improve delivery, reduce handling and transport costs, and reduce corruption, thereby bringing down the issue price substantially.
- Procurement of grain can be decentralized through creation of food grain banks in each village/block of the district, from where people can get subsidized foodgrains (including locally grown coarse cereals) through food coupons.

Secure Food Security for All

Food security is a physical, environmental, economic and social issue. It involves not just production, but access; not just output but process; not just technology but policy; not just global balance but also national conditions; not just national figures but household realities; not just rural but urban consumption; and not just quantity of food but also quality.

- The concept of food security should be broadened to make it holistic so as to mean "every individual has the physical, economic and environmental access to a balanced diet that includes the necessary macro and micro nutrients and safe drinking water, sanitation, environmental hygiene, primary health care and education so as to lead a healthy and productive life."
- Issues of food security are part of a bigger whole. Sustainable land and water management must be seen as directly linked to food security. Population growth, environmental sustainability, poverty reduction, agricultural production, distribution, marketing, credit and many other factors also need to be recognized as part of this whole.
- The major challenge is to produce additional food while conserving depleting natural resources. It is also to provide physical, economic and ecological access to food and nutrition security at the household level.
- Food security must focus on a diversified food basket, not food grains alone. Broad-based food security systems are not dependent on two-three species but on over 100 species that are underutilized. This will also prevent locally adapted grains from becoming extinct.
- Nutrition security must be given integrated attention by emphasizing

horticulture, animal husbandry, fishery, millets, pulses and several other resources for which India is traditionally known. There is need for investment in science and technology that will promote diversification.

- Food security must not be based on market, but rather on self-reliance and sufficiency. The approach should be one of moving from food security to food sovereignty.
- The elimination of hunger and malnutrition is not just a food problem. It is linked to poverty and population growth. Rising food output is essential but so are the slowing of population growth and maintaining the ecological balance.
- Food banks at grass root levels should be well maintained.

Capacity Building Mechanisms

In the 21st century, it is increasingly necessary, and increasingly feasible to take a whole systems approach to organized, positive change in rural places. For extension, that means helping farming people toward sustainably increasing productivity–particularly in the small-mixed farming systems in rain-fed areas, in upland areas, and in other places which have been neglected. It also demands measuring success in terms of the consumption of rural people, as well as of their production.

And that, in turn, will require agricultural extension systems which help farm men and women organize themselves in ways which empower them–to lead agricultural extension and to exert enough power and influence over agricultural research systems so that they generate useful, practical information which fits the needs and interests of those farming people.

- Agricultural extension must focus on increasing production and productivity of food and fiber in an economically and environmentally sustainable way. It must be done in a way which does not destroy rural livelihoods and rural communities.
- Extension activities should promote more comprehensive rural education and extension programmes directed particularly at rural poor, with major emphasis on efforts to reduce illiteracy, particularly among women and girls.
- It is important to organize education, extension and information, and skill empowerment on the basis of intensification, diversification and value addition of farming systems.
- Partnerships aimed at strengthening the knowledge base, and improving the dissemination of information, such as farm-to-farm technical assistance programmes, can help strengthen agri-extension. Public-private partnerships could be envisaged in basic sustainable agricultural techniques.
- Sustainable agriculture demonstration plots should be set up in

research stations, demonstration centres, seed production centres, farmers' training centres etc., of government, cooperative and non-government agencies.

- The system should support a new agricultural extension system that could meet the needs of information-hungry farmers, especially educated youth and women engaged in farming, and would empower them with new techniques and skills that foster sustainable agriculture.

Promote Awareness and Education Activities

National policies and planning should recognize that public awareness can play an important role in establishing a firm basis for sustainable agricultural resource conservation and use.

Public awareness should be considered in the development of all national programme activities:

- National strategies should identify objectives and strategies for public awareness, define target audiences, partners and tools for public outreach.
- Governments should recognize and encourage the work of NGOs in raising public awareness.
- Short-term and long-term courses specializing in sustainable agriculture should be run in agricultural universities. A separate curriculum for this subject should be developed for the regular graduate and postgraduate courses also.
- There is generally public apprehension about the possible consequences of agricultural biotechnology. It is important to organize public information programmes and public discussions that would help share relevant work of agricultural and biological scientists with the public.

Application of Research, Science and Technology

Scientific and technological development is not enough in itself. It is critical to make sure that sustainability and poverty reduction remain the guiding principles, and that we use our resources, harness our intellect, and direct our knowledge to benefit the poor, the hungry and the marginalized.

- The impact of agricultural research is decisive. Such research should, wherever possible, be coupled with on- farm activities in order that the context and purpose of the work are fully appreciated. Research should assist in the monitoring, evaluation, and improvement of on- farm efforts.
- Research should be undertaken in a participatory and collaborative manner to foster interaction and cooperation between rural people and research institutions. Other institutions must be involved appropriately whenever necessary.

- A comprehensive area-specific database of natural resources should be developed and made available for agriculture planning, implementation, research and extension. Existing data and information should be assembled, verified, and put in a usable and easily accessible form.
- Well-designed information technology packages should be developed, that could help serve as a market information network; weather, pest and disease monitoring system; and could be a storehouse of various current farming technologies and practices.
- Modern information technology should be used to reach the unreached. Educated youth must be attracted to and retained in farming through spreading science-based precision farming techniques, which are intellectually stimulating and economically rewarding.
- Training and capacity building should be undertaken in areas such as taxonomy, population biology, ethnobotany, and eco-regional and agro- ecological surveying. Specific research priorities need to be determined separately for each region.
- Additional research is necessary to develop drought tolerant, pest and disease-resistant crops, biological pest management, nitrogen fixation, more effective use of locally available organic materials, inter-cropping systems, and perennial crops, including agroforestry.
- Farming systems need to be designed so as to achieve the triple goals of more food, more income and more livelihoods per hectare of land. For this, it would be fruitful to harness the tools of eco-technologies resulting from a blend of traditional knowledge with frontier technologies. Such tools include biotechnology, information and communication technology, GIS mapping, space technology, renewable energy technologies (solar, wind, biomass, biogas), and management and marketing technologies.
- The revolution in biotechnology has both promises and problems. Biotechnology should be judiciously used so as to support the mission of environmental protection, poverty reduction and food security. But the adoption of every new technology must be accompanied by a precautionary package. In the case of biotechnology, biosafety and biosurviellance must be considered as important factors.
- Food should originate from efficient and environmentally benign production technologies that conserve and enhance the natural resource base of crop and animal husbandry, forestry, inland and marine fisheries.

ENVIRONMENTAL IMPACT OF AGRICULTURE

The environmental impact of agriculture varies based on the wide variety of agricultural practices employed around the world. Climate change and agriculture are interrelated processes, both of which take place on a global scale. Global warming is projected to have significant impacts on conditions affecting agriculture, including temperature, precipitation and glacial run-off.

These conditions determine the carrying capacity of the biosphere to produce enough food for the human population and domesticated animals. Rising carbon dioxide levels would also have effects, both detrimental and beneficial, on crop yields. The overall effect of climate change on agriculture will depend on the balance of these effects. Assessment of the effects of global climate changes on agriculture might help to properly anticipate and adapt farming to maximize agricultural production.

At the same time, agriculture has been shown to produce significant effects on climate change, primarily through the production and release of greenhouse gases such as carbon dioxide, methane, and nitrous oxide, but also by altering the Earth's land cover, which can change its ability to absorb or reflect heat and light, thus contributing to radiative forcing. Land use change such as deforestation and desertification, together with use of fossil fuels, are the major anthropogenic sources of carbon dioxide; agriculture itself is the major contributor to increasing methane and nitrous oxide concentrations in earth's atmosphere.

CONCEPTUALIZATION OF TECHNOLOGICAL CHANGE IN AGRICULTURE

Technological change is among a small handful of topics in what is now referred to as the sociology of agriculture — along with part-time farming and analysis of the farm family — that have received attention during all three major phases in rural sociology. From the 1930s through the 1950s there were several major studies by sociologists of the impacts of technological change, most of which centered on the rapid mechanization of southern agriculture that began during the Great Depression.

The concern of many of the early rural sociologists with the socioeconomic consequences of technological change, however, largely lapsed during the succeeding phase of rural sociology in which the diffusion-adoption perspective was dominant. In the era of the "new sociology of agriculture," we again see a rekindling of concern with the social consequences of technological change. Moreover, interest in the consequences of technological change has become expressed in more social structural terms and, as demonstrated later, has become more closely integrated with analyses of the genesis of new agricultural technologies.

Several other important observations can be made on the rural sociological literature on technological change in agriculture over the past decade. First, mechanization has received the lion's share of attention; technological change in the forms of improved varieties of crops and breeds of livestock or of agricultural chemicals has been the focus of only a few major ex post studies. Second, as Berardi has observed, there are many problems in decomposing the effects of technological change on farm structure from the effects of other variables such as public policy or changing wage rates; therefore, rural sociologists have often relied heavily on the primary research of others (*e.g.*, agricultural economists and geographers with complementary expertise in assessing techno logical impacts) when developing comprehensive assessments of the socioeconomic impacts of technological change.

Third, the conceptualization of technological change in agriculture by rural sociologists has been heavily influenced by the work of the agricultural economist, Willard Cochrane, who developed a theory of the "treadmill of technology" in the late 1950s. Briefly, Cochrane's theory of technological change builds on a key observation from rural sociological research on the diffusion-adoption of innovations-that large farmers who are entrepreneurial and nonrisk-averse tend to be the first to adopt new agricultural technologies.

By adopting early, these farmers enjoy "innovators' rents" because of their ability to reduce their average per unit production costs. Cochrane posits that after these innovators and early adopters have employed the new agricultural technology, aggregate output of the particular commodity involved will begin to increase and prices will decline disproportionately (because agricultural commodities tend to have a low price elasticity of demand).

Declining prices then begin to force nonadopters to utilize the technology. These subsequent adopters gain very little from the technology; they adopt it merely to be able to stay in business. Slow adopters or nonadopters will tend to be forced out of agriculture because they can no longer compete due to their high average costs of production. Cochrane considers the technological change process to be a "treadmill" in that the majority of farmers, who receive little or no benefit from the technology, are nonetheless forced to continue to adopt new agricultural technologies in order to stay in business.

Cochrane has also observed that the benefits of technological change tend over the long term to be realized in the form of rising land values, particularly if there are commodity programs in effect to cushion a commodity sector from declining crop prices caused by increases in output. The benefits of technological change (innovators' rents from early adoption) tend to be reinvested in farm land, causing increases in land prices and benefiting large landowners over smaller property owners and tenants. These effects, in Cochrane's view, are heightened when commodity programs support crop prices, thereby increasing the level of innovators' rents and intensifying the extent to which these benefits are capitalized in asset values.

Cochrane's theory of the treadmill of technology has been widely accepted in recent rural sociological work as an orienting perspective because it has effectively linked knowledge from diverse origins (rural sociological research on diffusion-adoption, agricultural economics research on technological change, and so on). Cochrane's theory, however, has its limits — in particular, its inapplicability to industrial agriculture, as noted by LeVeen.

Moreover, the notion of the treadmill of technology gives little attention to the origins of technology, that is, to why technologies that have displaced labor or benefited larger operators over smaller ones have been developed, a topic that, has recently come to be a major research issue for those working in the sociology of agricultural science. Nonetheless, with the exception of sociologists of agriculture who have focused on industrial agriculture, rural sociologists have yet to develop a distinctly sociological theory of technological change and its socioeconomic consequences.

While rural sociologists have not yet developed a comprehensive and distinctly sociological theory of technological change and its consequences applicable to both family farming and industrial agriculture, the new sociology of agriculture as applied to technological change has made some significant advances. Several of the most important of these advances are summarized below.

Among the most significant research papers on the topic of technological change in agriculture were two published in the late 1970s that dealt with limitations of the classical diffusion-adoption model. The first, by Pampel and van Es, was an otherwise traditional adoption study that included, in addition to "commercial-profitable" innovations that were the major focus of previous diffusion-adoption studies, a set of "environmental innovations" such as contour farming, terracing, sod waterways, and use of rotation cover and reduced tillage. Pampel and van Es discovered that the correlates of adoption of environmental innovations were far different than those of commercial innovations. For example, while large-scale farmers were observed to be the earliest adopters of commercial technologies, there was no significant relationship between farm size and the adoption of environmental innovations.

They concluded their study by cautioning rural sociologists that the diffusion-adoption tradition, because it had ignored dimensions of innovative behaviour such as the adoption of environmental-conservation technologies, "may have provided within rural sociology a field of knowledge with a narrow empirical foundation on which to base its generalizations". They also emphasized that given the tendency of most environmental innovations to be unprofitable, it was not clear that voluntary strategies to achieve soil conservation and related environmental goals would be successful.

Writing two years after the Pampel and van Es study, Goss focused more systematically on the limitations of diffusion-adoption research. The principal failings of diffusion-adoption research, in Goss' view, were (1) that it was uncritical, if not promotional, towards patterns of technological change that may be detrimental to the interests of society and/or farmers, and (2) that the focusing of research effort on the adoption component of the technology change process had caused rural sociologists to ignore the consequences of technological change for various groups in society.

Goss summarized the early 1970s literature critical of the Green Revolution and pointed out that the Green Revolution had been initiated with the same promotional orientation towards technology that had predominated during the heyday of the diffusion adoption tradition in U.S. rural sociology. He noted that many social scientists who had once worked within the diffusion-adoption tradition in international agricultural activities had begun to recognize the limits of the adoption paradigm.

He argued, however, that this point had not been well recognized in domestic rural sociology and that rural sociology must begin to develop a more comprehensive programme of research on the consequences of technological change in agriculture.

In a manner similar to Goss' exegesis of the Green Revolution literature in terms of the impacts of technological change on social groups and classes in agriculture, Stockdale, in an influential article, raised comparable issues in terms of the ecological aspects of modern agricultural technology. Noting that the contribution of technological change to American agricultural productivity had often been held up as a model for other nations to emulate, Stockdale argued that social scientists had tended to ignore the adverse ecological impacts of energy-intensive technology in the United States and elsewhere. Like Goss, Stockdale called for an expanded programme of research on technological change in agriculture that would emphasize the social and ecological consequences of new agricultural technologies.

Among the most innovative thrusts in sociological research on technological change in agriculture have been the series of books, monographs, and articles published by Friedland and associates. Friedland and his colleagues have focused on technological change in industrial agriculture in California and have identified a number of components of the technological change process that have tended to be ignored in the earlier rural sociological literature.

As noted earlier, Friedland has focused on several commodity systems (chiefly tomatoes, lettuce, and grapes) and has found that the social organization of each commodity system — and hence the structural context of technological change — is distinctive.

Nonetheless, Friedland and his associates have demonstrated that the technological change process in industrial agriculture generally is heavily influenced by considerations of the cost of, access to, and control over labor and by the relationships of grower organizations with public and private components of the agricultural research system. They have also demonstrated that the nature of the technological change process and its consequences depend upon the structure of the commodity system. For example, in the aftermath of the termination of the bracero programme in 1964, a mechanical tomato harvester technology was widely adopted in California, displacing three-quarters of California tomato producers and the bulk of their agricultural wage labor forces. In lettuce, however, grower-shippers were able to utilize the threat of lettuce harvest mechanization to discipline the labor force and to avoid the costly process of harvest mechanization.

A further component of the recent rural sociological literature on technological change, one to which Friedland and Friedland and associates have made major contributions, is social impact assessment of new agricultural technologies. Social impact assessments of new agricultural technologies are a distinctive type of research in that they often tend to rely primarily on secondary data or on the results from the published literature rather than on primary research and data collection; this is particularly the case if the assessment must be prepared over a short period of time.

Other distinctive aspects of social impact assessment of new agricultural technologies are, first, that the assessment is ex ante (rather than ex post, as in most previous rural sociological research on technological change) and, second, that the orientation towards these emerging technologies tends to be questioning and skeptical, rather than accepting and promotional. Rural sociologists who have conducted social impact assessments of new technologies have approached their subject matter from the vantage point that most new technologies tend to involve an unequal distribution of costs and benefits and that advance projections of these patterns of the distribution of costs and benefits can enable social scientists and affected groups to redress these distributional impacts more effectively.

An example of the type of knowledge base that has been developed, in part, as a building block for social-impact assessment research is Berardi's comprehensive review of the impacts of agricultural mechanization. Berardi reviewed the re sults of over 180 studies of agricultural mechanization in the United States and elsewhere and summarized the major empirical regularities in studies of a variety of types of mechanization in a diversity of commodity systems. She concluded by pointing out that the rich literature on agricultural mechanization — most of it prepared by nonsociologists — is sufficient so that sociologists may contemplate conducting rigorous ex ante studies of the impacts of new mechanical technologies in agriculture.

Unfortunately, as noted earlier, sociological knowledge on the socioeconomic impacts of mechanization is far greater than that on "biochemical" technologies (plant varieties, fertilizers, insecticides, herbicides, and so on). There have only recently been sociological studies that have delivered on Stockdale's and Berardi's admonitions in the area of biochemical technologies in plant and animal agriculture. It is likely, however, that the great attention that has already been paid by rural sociologists to emerging biotechnologies in agriculture will result in a sustained research programme into the socioeconomic consequences of these new technologies. It should be noted in closing, however, that while there is great merit in inventorying the results of existing studies on various facets of technological change — as, for example, Berardi has done — sociologists should exercise caution, as technological change should not be assumed to have invariant consequences across time and space. In particular, as noted above, there have been some major recent changes in the structure of U.S. agriculture that one might expect will have significant effects on the technological change process and its consequences.

For example, instead of technological change tending to result in the decline of small farms, as it did during the bulk of the post-World War II period, it may be quite likely that new agricultural technologies such as biotechnology-derived plant varieties, growth regulators, and bovine growth hormone may tend to result in the disproportionate demise of medium-sized, full-time family farmers. Thus future rural sociological research on technological change in agriculture must be cognizant of ongoing changes in the structure of agriculture and how these changes affect the types of agricultural technologies that are employed, the processes of technological change, and their consequences.

AGRICULTURE AND THE RURAL COMMUNITY

Perhaps in no other area of the "new sociology of agriculture" is there so much apparent continuity with the pre-Korean War tradition of rural sociological research as in research on farm structure and rural communities. The early rural sociologists primarily conceived of agriculture as one of the major institutional ensembles affecting rural community life and, especially during the 1930s, devoted much of their research effort relating to agriculture to understanding the implications of agricultural structural changes for rural community life and viability.

The conceptualization of agriculture and rural community interactions and the methodologies for exploring these interactions in pre-Korean War and post-1970 rural sociological research, however, exhibit dramatic differences. Whereas most of the early rural sociologists tended to be less interested in agriculture per se and were principally concerned with understanding rural community institutions, contemporary researchers have tended to be more interested in structural change in agriculture and have deemphasized understanding the totality of the fabric of rural social life. The research tools and data sources of these two generations of rural sociological researchers are also dramatically different.

The early rural sociologists relied on rural social surveys, while the bulk of contemporary research is based on county-level census data. A further difference between the early and contemporary rural sociologists of agriculture in their work relating to rural communities concerns the pivot of the latter studies around Goldschmidt's "As You Sow". This is somewhat ironic in that Goldschmidt's study was anthropological and qualitative, whereas the research conducted since 1970 that has drawn on Goldschmidt's work has largely been quantitative and has involved far less "anthropological" detail on rural communities than did Goldschmidt's.

Rodefeld is generally credited with the "rediscovery" and popularization of Goldschmidt's pioneering work among rural sociologists, though Heffernan made important contributions to elevating "Goldschmidt-style" research to a prominent position in rural sociology. Nonetheless, after Rodefeld's refocusing of rural sociological attention on the Goldschmidt thesis, Goldschmidt's book has been virtually an obligatory citation in research articles on this topic. Despite some trenchant criticisms of Goldschmidt's perspective and research methodology, Goldschmidt's general hypotheses remain accepted theoretically and thus continue to shape, if not dominate, research on farm structure and rural communities within the "new sociology of agriculture."

Briefly, Goldschmidt's thesis was that there are differences in the quality of rural community life depending upon the social organization of agriculture and its attendant occupational structure. Goldschmidt observed that the quality of rural life in two carefully matched California communities in the San Joaquin Valley differed dramatically and that these differences were largely accounted for by the organization of agriculture.

He argued that Dinuba, a community surrounded by "family" farms, had a greater number of businesses, a larger retail sales volume, a higher per capita income, and a greater diversity of social, educational, recreational, and cultural institutions than did Arvin, a community surrounded by large, industrial, corporate farms. These differences were principally accounted for by an intervening variable—community occupational composition—that was seen to be strongly influenced by the social organization of agriculture and which directly contributed to the dramatic variations in indicators of the quality of life.

Goldschmidt reported that the high ratio of laborers to other occupational groups in Arvin accounted for its low level of community viability relative to Dinuba. From these observations Goldschmidt inferred that the rise of large-scale corporate agriculture then underway in California and elsewhere would have adverse implications for rural communities.

Goldschmidt's study received essentially no further attention until it was referred to in a 1968 hearing on corporation farming before the Senate Select Committee on Small Business. LaRose and Sonka and Heady soon reported further data consistent with the Goldschmidt study. In addition to Rodefeld's and Heffernan's efforts to resurrect Goldschmidt's work in rural sociological circles, there was a major research effort by the Small Farm Viability Project in California, drawing largely on Isao Fujimoto's unpublished research, which presented a larger restudy of the Goldschmidt thesis. The results largely supported Goldschmidt's findings of over three decades earlier. Flora and Conboy also reported results consistent with the thesis.

Since that time there have been numerous studies in the new sociology of agriculture tradition that have explored the Goldschmidt thesis in a number of contexts.

Several observations can be made on the conduct of recent research in the Goldschmidt tradition, what we might call "neoGoldschmidt" research. First, it must be stressed that contemporary rural sociological researchers have taken considerable liberties with Goldschmidt's formulations. As noted earlier, Goldschmidt's method was a comparative anthropological case study of two communities and eschewed quantitative analysis of the sort that has predominated in the subsequent rural sociological literature. Also, Goldschmidt's analysis focused on several aspects of the property relations and control over labour, such as absentee ownership and use of ethnic minority laborers, rather than farm size, which has been the most important farm structural indicator in the neo-Goldschmidt tradition. In that regard it is important to stress that the "family" farms of Dinuba were by no means "small" farms; in fact, the family farms of Dinuba were generally larger in gross sales terms than most large farms in the Midwest today.

Second, neo- Goldschmidt research has largely focused on a single direction of causality—the impact of agricultural structure on rural communities—and has largely ignored other types of interrelations between agriculture and rural communities. Third, most studies have been based on Census of Agriculture data and have employed countylevel data as a proxy for community characteristics, though some researchers have utilized survey data to infer community quality of life consequences of changing agricultural structures. Fourth, despite obvious major regional variations in farm structures and in rural communities and their economies, there has, until very recently, been a virtual lack of a comparative perspective on farm and community structures.

Goldschmidt and Harris and Gilbert are partial exceptions to this generalization, since they have done research using states as the unit of analysis, though this procedure compounds the aggregation bias problems. Very recently, nonetheless, there has appeared a series of regionally based studies of the effects of farm structure on rural community viability and quality of life commissioned by the Office of Technology Assessment. These studies are Buttel et al., Skees and Swanson, Flora and Flora, MacCannell, and van Es et al. as published in abridged form in Swanson. These five studies were very briefly summarized in the main report. Together they indicate strongly that there are pronounced regional variations in how changing farm structures affect rural communities. The impacts of farm structural changes on rural communities are generally most pronounced in the Great Plains and industrial agriculture states (California, Arizona, Texas, and Florida), less so in the South and Midwest, and very little in the Northeast. There is evidence that the degree to which farm structure affects rural communities is directly associated with the dependence of the rural community on agriculture.

The general model pursued in these neo- Goldschmidt studies has been to estimate coefficients for a causal chain in which farm structural change indicators are seen to affect the composition of the farm population (e.g., as measured by indicators of the size of the farm population, the proportion of full-time hired workers) and the size and composition of the rural population. These clusters of variables are then seen to affect indicators of the quality of life in rural-agricultural communities (e.g., proportion of families in poverty, median family income, per capital retail sales, unemployment rate).

This general model is sometimes supplemented by technological factors as antecedents of farm structure, since technological change is considered to be one of the major factors influencing farm structural change. More recently, studies in this genre have gone beyond the cross-sectional census data designs that predominated from the mid-1970s to the early 1980s and have included measures of change, where appropriate, in variables in the model. One limit to studies that use census and other data based on single-year assessments of farm income is that net farm incomes are highly variable over the short-run; for example, as many as 20 percent of farms with annual sales of over $250,000 had poverty-level incomes in 1986.

Heffernan has effectively summarized the general configuration of results in the neo-Goldschmidt literature by noting that "all relevant research to date suggests that a corporate type of agriculture results in a reduction in the quality of [community] life for at least some people, especially the hired workers in rural communities." As such, the results continue to lend support to the Goldschmidt thesis.

But there have recently been several expressions of dissatisfaction with Goldschmidt's perspective and with the generalizations that have been made from research in this tradition. A major theoretical concern is that Goldschmidt and many researchers working in this tradition have taken large-scale and corporate agriculture to be basically synonymous when, in fact, not all large-scale farms are corporate or industrial-type farms. Much more important than industrial-type farms in farm numbers and shares of sales, assets, and profits are larger-than-family farms.

Empirically, as Goldschmidt-style research has been done in the Midwest and other traditional family farming areas, there have been admonitions that large-scale agriculture, especially of a nonindustrial character, does not necessarily lead to adverse impacts on rural communities, though some studies note that increased scale of production could have detrimental impacts on community services. A further area of dissatisfaction with research on farm structure and rural communities is that the treatment of farm structure in this literature has generally failed to incorporate new concepts relating to agricultural dualism. Typically farm size or scale is operationalized as a single variable representing a central tendency measure of farm size, which was a suitable procedure when there was a singular process of differentiation among farms. But with the emerging dualism in U.S. farm structure since the late 1970s, there has been little change in mean acres per farm since the 1969 census. This measure of central tendency thus ignores the processes of dualism that have contributed to increased concentration with apparent stability in average acres per farm since the late 1960s.

The neo- Goldschmidt conceptualization of an agricultural structure in which particular communities can be characterized along a continuum of farm sizes ranging from small to large has also been questioned by Wimberley who, using multiple indicators from county-level Census of Agriculture data and a longitudinal design, found that there are instead three continua relating to degree of "corporate-commercial" agriculture, "largefarm-area" agriculture, and "small-farm" agriculture. Using Wimberley's measures and including sources of off-farm employment as well, Reif found that large-farm-area agriculture and core-industry structures were related both to family poverty and income inequality in counties.

Thus, while studies of the impacts of changing farm structures on rural communities have been a productive area of research in the new sociology agriculture, a number of research frontiers for the future are suggested by recent criticism. There is a need to diversify the methodological approach, consider alternative directions of causality, introduce more sophistication in terms of the measurement of change in farm structure, and expand recent efforts at comparative regional and interdisciplinary research.

The work of Vail represents an encouraging thrust in this direction. Vail has conducted extensive research on the relationships between agriculture and rural communities in Maine in which a variety of factors of demonstrated importance—the reciprocal relations between agriculture and rural communities, the role of an increasingly dualistic agrarian structure, nonfarm aspects of agriculture, and the historically specific nature of farm-community interactions—have been incorporated into a single study.

The Farm Crisis Of The 1980s

The notion that American agriculture is no longer (if it ever was) isolated from the remainder of the political economy was driven home clearly by the farm crisis of the 1980s. This crisis was brought by the conjunction of a number of factors in the 1970s, including: accelerated demand for grain on world markets, public policies that encouraged massive capital investments and expansion of production, and a political-economic context that encouraged land speculation. The farm crisis emerged in the 1980s for a number of reasons. Global agricultural export markets shrank, in large part due to the global recession. Farm commodity prices and ultimately net farm incomes (exclusive of federal commodity programme payments and price supports) declined due to the glutting of world grain markets. Reagan administration fiscal and monetary policies led to high real interest rates, which were especially detrimental to agriculture because of its interest rate sensitivity; accordingly, debt-asset ratios among farmers increased rapidly.

The 1980s also witnessed the emergence of new Third World agricultural export competitors (e.g., Argentina, Brazil) because of the global debt crisis and these countries' need to increase exports in order to repay foreign loans. Further, federal fiscal austerity eroded the federal government's ability to maintain farm incomes through commodity programmes (even though commodity programme expenditures soared during the decade). The result was a massive decapitalization of American agriculture, especially in the Midwest and Great Plains, as farmland prices in the states most affected by the crisis declined by over 50 percent.

This crisis attracted a good deal of attention by the mass media and by agricultural economists. It was not, however, until the middle of the decade that much was published about the farm crisis by rural sociologists.

Most of this research can be divided into four categories:

1. Neo-Marxian (and related neo-Weberian) analyses of the continuing penetration of capitalism into and the role of the state in agriculture,
2. Analyses of the differences in the characteristics between farmers that were and were not experiencing "financial stress,"
3. Studies of the effects of farm financial crisis and failure on the farm families whose enterprises were affected,
4. Studies of public reactions to the farm crisis.

The neo-Marxian (and neo-Weberian) approaches have largely focused on understanding continuing cycles of crises in agriculture under capitalism. Examples of this genre specifically focused on the crisis of the 1980s include Berlan, Bonanno, Mooney, McIntosh and Zey-Ferrell, Zey-Ferrell and McIntosh, Goodman and Redclift, and Buttel.

Studies in the second category, analyses of the characteristics of farms with various levels of "farm financial stress," deal more with the effects of the political and economic environment of agriculture on farms of different characteristics. These studies have produced inconsistent results due to regional, ethnic, and other variations within the farming population. For example, farm-size variables were found to be directly correlated with financial difficulty on Iowa farms and unrelated for Texas and North Dakota farms.

Yet aggregate national data revealed that larger farms were more profitable, a relationship that obscured the underlying failure of many middlesize farms. Farm families with fewer years of farming and younger ages were affected more severely. However, Murdock et al. found the cohort that had started farming between the end of World War II and the mid1960s also had higher rates of financial difficulties, suggesting a life-cycle effect for those farm families bringing another generation into the farm. The literature does, however, show agreement on the relationship between other variables and financial crisis. For example, higher debt-to-asset ratios brought by low commodity prices in a time of deflation of asset values were associated with financial problems.

The third category, analysis of the effects of the farm financial crisis and failure on the farm families whose enterprises were affected directly, has largely been the work of William and Judith Heffernan, who have conducted in-depth interviews among Missouri farm families who were in severe financial crisis or had been displaced from agriculture. These families experienced a host of difficulties beyond the simple loss of their farms and the disorientation that accompanies the loss of a valued way of life. Among their major results are the following: Many farm families felt socially isolated and were not well-prepared for participation in nonfarm labour markets.

Local employment opportunities were often unavailable, and tax laws devastated some families economically after they had attempted to leave farming with some resources with which to start a new life. While the long-term effects of farm failure for these families were not yet known, it was clear that some families were adapting to their new situations successfully while others were at risk of suffering long-term poverty. Graham's study of displaced farm families in New York State bears out these findings. This category of studies has implications for the mutual effects of agriculture and its environment as the displaced farm families affect local labour markets and community social structures.

The final category of research on the 1980s farm crisis, on public reactions to the crisis, is best represented by Lyson's research on public apathy. Drawing on theories of public opinion formation, Lyson demonstrates that several factors led to a surprisingly high degree of apathy by nonfarm people toward the farm crisis, despite the massive media attention the crisis attracted. He notes that there has been a trend toward farmers and consumers having very little direct contact and that the relationship between food prices and farm income has become concealed. Most importantly, Americans have come to view "bad news" from the farm sector—the cost-price squeeze, high debt loads, and the loss of family farms—as being normal. Thus the American public, though "interested and concerned" about the economic woes of farmers, exhibited apathy toward the farm crisis.

A major neglected area in the rural sociology literature on the farm crisis to date is studies of the effects of the farm crisis on rural communities. This topic pertains, in particular, to how conditions in agriculture affect its environment, especially at the local level. Financial losses to local agribusiness firms when farms fail, losses of farm families in local rural market areas, declining property tax revenues due to deflated farm asset values, and other problems associated with the declining prosperity of agriculture have likely had severe effects on local services and on the local social fabric.

THE AGRICULTURAL SCIENCE

We noted several times in the preceding that the era of the new sociology of agriculture was initiated in the aftermath of a distinctly new political milieu of agricultural science and research. Hightower *Hard Tomatoes, Hard Times,* although sensational in tone and misleading on several aspects of the public agricultural research system, attracted a great deal of attention from defenders and critics of the land-grant research system. Most importantly, Hightower's book (along with contemporary ecological criticism of U.S. agricultural technology) opened up several new lines of social science inquiry into the structure and functioning of the public agricultural research system.

Probably in no other state did public agricultural science become as controversial as in California after the publication of Hightower's book. The California Agricultural Experiment Station has long had close relationships with industrial farmers and other segments of agribusiness and had been held directly responsible in the eyes of many for developing new technologies, such as the tomato harvester, which had resulted in the rapid loss of tomato farmers and the displacement of hired farm workers. It was thus not surprising that the initial sociological inquiries into the conduct of public agricultural science were conducted in California.

The first set of studies by Fujimoto and others at the University of California, Davis, were directly inspired by Hightower and were largely focused on how agricultural scientists made research problem choices. In particular, Fujimoto and colleagues devoted attention to how research problem choices were determined by "internal" factors (such as tenure and promotion considerations) and "external" factors (such as the influence of agribusiness interests and grants).

They found on the basis of personal interviews with scientists that external factors were quite important, though internal factors also exerted major influence on scientists' research priorities. The second set of studies focused on California was initiated by Friedland and associates, who examined the processes by which agricultural commodity groups organized to exert influence on individual scientists, departments, and the University of California system as a whole in order to ensure the development and availability of technologies that would enhance their profitability.

Friedland and associates gave particular attention to how the responses of University of California research officials and scientists came to emphasize mechanization and mechanization-related technology that enabled growers to displace farm workers or to enhance the control of growers over their hired labour force.

Based on the results of several years of research during the 1970s, Friedland and Kappel published a short, but still influential, monograph calling attention to biases in the public agricultural research system and outlining mechanisms by which social scientists and groups adversely affected by particular types of agricultural research could achieve changes in the land-grant system.

Shortly after the California studies were initiated, a research group at the University of Kentucky led by William B. Lacy and Lawrence Busch began a programme of inquiry into the sociology of agricultural science that continues to make major contributions to a sociological understanding of the processes by which new agricultural technologies are generated. The first publication of this group, by Busch, was a theoretical critique of the prevailing conception of scientific knowledge by researchers in the diffusionadoption tradition. Subsequent publications by the Lacy-Busch group have focused on a variety of aspects of the structure and functioning of public agricultural research activities.

Two books published by the Lacy-Busch group have had particular impact on rural sociology and the larger social science community exploring agricultural research. The first, *Science and Agricultural Development*, edited by Busch and containing several papers by his research colleagues, was the first book-length treatment of agricultural research in the tradition of the new sociology of agriculture.

The chapters in the Busch anthology most germane to the sociology of agricultural science are the following: Randolph and Sachs compared the establishment of medical and agricultural science, emphasizing how a reductionist orientation and technocratic ideology had been developed in each during the process of institutionalization.

Busch and Lacy summarized the existing literature and their preliminary research on sources of influence in research problem choices in the agricultural sciences. Striving to further the social scientific conceptualization of agricultural science and technology as social products, they explored two rival social science views of the social construction and social structure of science. One, the "internalist" view deriving from Merton work, Sociology of Science, essentially holds that the outcomes of the scientific research process are determined by the "norms of science," held by practitioners and reinforced within scientific institutions, and by the very subject matter itself.

The other perspective, the "externalist" view, argues that the outcomes of the research process are largely shaped by the institutions, interests, and availability of resources that originate and are determined outside of scientific disciplines and research institutions. They suggest that both perspectives have a certain validity: "[T]he internal and external positions are best viewed as distinct strata in the life-worlds of practicing scientists. Thus, for each discipline and institution, science is a continuous process of multilevel negotiation regarding the sources of influence on research". A final paper in the Busch collection of particular relevance to the new sociology of agriculture was Busch and Sachs', which employed a world-systems framework derived from Wallerstein to examine the establishment of the agricultural sciences in the First and Third Worlds since the seventeenth century.

Busch and Lacy are known best for their book, *Science, Agri-culture, and the Politics of Research* culture, and the Politics of Research, which summarized an ambitious study of research problem choice through the use of historical methods, personal interviews, and sample survey data. Following several introductory chapters that discuss the establishment of public agricultural research, the organization of the public agricultural research system, the social demography of agricultural scientists, and the structure of communication in the agricultural sciences, they explore three major sources of influences on agricultural research: disciplinary, organizational, and extra-organizational.

Busch and Lacy found scientific disciplines and disciplinary organizations to be powerful and growing influences over scientists' research problem choices. Scientific disciplines affect scientists' research choices in a variety of ways, particularly through controlling peer-reviewed journals and establishing the norms adopted by land-grant universities and the Agricultural Research Service (ARS) in tenure and promotion decisions. Busch and Lacy also found dramatic differences among the agricultural science disciplines according to the degree to which they emphasize basic or applied research, the rate at which they accept papers in scientific journals, their norms for scientific productivity, and the degree to which scientists exhibit consensus on what are "hot" topics in the field.

Organizational context—including ARS vs. land-grant university context and the institutional prestige of the land-grant institution—was found to be a significant influence on research problem choices. Busch and Lacy found that ARS scientists were more highly constrained in their problem choices than land-grant agricultural scientists.

They also found that institutional prestige was positively related to the degree to which extra-organizational influences and constraints on research choices were manifest. Among the extra-organizational influences, granting agencies and organizations (those hiring scientists on a short-term consultancy basis) were the most direct and pervasive. Whereas there were very large differences among the disciplines in the extent to which their scientists relied heavily on extramural support, Busch and Lacy emphasize that fiscal constraints on public research appropriations are increasing the pressures on scientists to obtain grants and contracts from government agencies, agribusiness firms, foundations, and so on; accordingly, scientists are becoming less autonomous in their research problem choices.

Until the early 1980s, only a small community of rural sociologists was interested in the sociology of agricultural science. This was to change dramatically, however, in the aftermath of a workshop on biotechnology in the summer of 1982 at the University of Kentucky organized by B. Koppel, Busch, and Lacy. Since that conference there has been growing attention to a variety of facets of biotechnology as key focal points of research in the sociology of agricultural science.

This research on biotechnology now represents a large literature, and the approaches developed therein have begun to substantially influence how rural sociologists are exploring other types of agricultural technology such as information technologies. This new line of research has been salutary in yet another respect; sociologists interested in biotechnology have begun to make major progress in uniting the sociology of agricultural science with sociological research on technological change and social impact assessment of new agricultural technologies.

The wide-ranging and potentially integrative character of research on biotechnology, agricultural research, and agricultural change is amply illustrated by rural sociologists having given major attention to no less than 10 major themes under this rubric. With regard to the first, the politics of the evolution of the public agricultural research system, Buttel and Kenney and Kloppenburg found that the basic science discoveries (especially recombinant DNA) relating to biotechnology, the promise of biotechnology to agribusiness firms for developing new input commodities, and federal government concern with the U.S. competitive position in the "high technologies" have rekindled scrutiny of the land-grant and ARS systems for placing too little emphasis on basic biological research.

The result was a changing political climate of public agricultural research that has led to rapid expansion of biotechnology programmes in the public agricultural research institutions and to a deemphasis on traditional, applied research.

These changes in the research emphases of public agricultural research institutions have stimulated a second area of research in the sociology of biotechnology: industry-university relationships and the changing division of labour between public and private research. Buttel et al., for example, have found that industrial funding of agricultural research at the land-grant universities has increased significantly and this has generated new concerns that agribusiness firms are acquiring excessive influence on research priorities in public research institutions.

Lacy and Busch have identified a large number of likely impacts of closer industry-university relationships in agricultural biotechnology: further decline of public plant breeding (especially varietal releases), greater concentration of scientific talent at a small number of land-grant universities, decreased autonomy of public researchers in establishing their own research agendas, increased secrecy, and greater restrictions on the free flow of scientific information.

A third area of research relating to biotechnology has been on agribusiness reorganization. Kenney, Kloppenburg, and Buttel et al. have noted that the rise of biotechnology research has reinforced or led directly to major structural reorganization in the agribusiness industries. With the passage of the Plant Variety Protection Act of 1970, the ability of private firms to acquire patent-like protection for new crop varieties increased the perceived profit potential of seed companies, leading to the incorporation of most major independent seed companies into multinational corporations as wholly owned subsidiaries or joint-venture partners.

Biotechnology reinforced this trend, since seeds are projected to become the major commercial vehicle for plant biotechnology; moreover, the basic sciences underlying biotechnology (molecular and cell biology) are quite similar to those that will ultimately be applied in other areas of agricultural inputs such as agricultural chemicals, innoculants, and animal pharmaceuticals. Thus biotechnology has increased the benefits of uniting chemical and plant breeding product lines within integrated multinational firms that manufacture several agricultural input products. Kenney has given particular attention to the role of small venture-capital or start-up companies in biotechnology and has analyzed the processes by which they have developed linkages with large companies.

Goodman et al. have stressed that biotechnology will have two distinct applications in agriculture—"appropriationism" and "substitutionism"—with different consequences for structural reorganization of the food and agricultural industries. By "appropriationism" Goodman et al. mean the use of biotechnology to develop new inputs, such as new crop varieties and animal breeds, for agricultural production. Appropriationist applications of agricultural biotechnology will tend to result in the integration of agricultural input manufacturing with comparable product lines in other life science industries such as chemicals and pharmaceuticals. "Substitutionism," on the other hand, refers to the use of biotechnology to substitute for existing products.

Substitutionism will therefore have the effect of severing existing production "chains," in which specific raw materials are employed to produce particular final products (e.g., substituting single-cell proteins for vegetable proteins as animal feed supplements). The implications of "substitutionist" applications of biotechnology will also include shifting food and fiber production from the farm to the factory (e.g., substituting industrial tissue-cultured thaumatin for cane sugar). Substitutionist applications will tend to lead to integration of the food and fiber processing and chemical industries.

Fourth, several researchers have examined how biotechnology has affected and will continue to affect plant breeding, including but not limited to public-sector plant breeding. Kloppenburg has conducted an extensive historical analysis of plant breeding and the seed industry in the United States.

He has argued that the trends under way as a direct and indirect result of biotechnology promise to increase the penetration of agribusiness interests into land-grant university plant breeding programmes and to complete the shift, under way since the late 1930s, to curtail the role of land-grant plant breeders in developing public-domain finished plant varieties.

Hansen et al. have reached similar conclusions on the basis of interviews with plant breeders and biotechnology researchers across the public and private sectors. Hansen et al. have noted as well that the increased importance of biotechnology in public-sector plant breeding programmes, combined with new avenues for agribusiness influence, will tend to result in the demise of public plant breeding efforts for minor crops.

A fifth area of inquiry in biotechnology, agriculture, and agricultural research has been on the role of intellectual property restrictions and other proprietary considerations in affecting research priorities and the nature of the new technologies that are developed. Kloppenburg, for example, has called attention to private-sector plant breeding firms' longstanding concern about the perceived inadequacy of intellectual property institutions to protect their investments in plant breeding. In the absence of such restrictions for sexually reproduced crops prior to 1970, most seed companies focused primarily on hybrid seeds (largely hybrid corn), which, because of their reproductive instability, afforded a "natural," biological protection against farmers saving their own seeds and circumventing annual purchases of seeds from private companies.

Kloppenburg has noted, however, that the elaboration of new legal means of privatizing plant research has increased corporate interest in plant biotechnology. Kloppenburg, Kenney, and Hansen et al. have suggested that the role of intellectual property restrictions, combined with the fact that most agricultural input firms have both seed and chemical product lines, may lead to the development of "seed-chemical packages" that tie the farmer more directly to purchased, capital intensive inputs.

Sixth, Kloppenburg and Buttel have begun to explore the relationships between the public and private sectors in terms of recently developed "theories of the state". They have begun this task by conceptualizing public agricultural research and government commodity programmes as reflecting contradictions of state intervention.

They note public agricultural research with state intervention has tended historically to be expanded during times of socioeconomic crisis. Agricultural research, however, can be seen to set the stage for future crises in the agricultural economy by creating long-term increases in agricultural productivity.

They discuss the emergence and resolution of these contradictions in the context of the Great Depression and the current era of biotechnology and farm crisis.

A seventh area of research pursued within the rubric of biotechnology has been the implications of this new trajectory of research for genetic resources. Kloppenburg and Hansen et al. have argued that the increased role of intellectual property restrictions combined with the new opportunities for private firms in developing new purchased input commodities will tend to lead to a decline in germplasm flows between the public and private sectors. In the absence of greatly expanded public and quasi-public appropriations for germplasm preservation, these trends are seen to threaten the genetic resource base and perhaps to constrain plant improvement in the future. Kloppenburg and Kleinman have also examined the forces, including the rise of biotechnology, that have led to North-South conflict over control over crop germplasm.

Eighth, as noted earlier, considerable attention has been focused on how new biotechnology-related inputs such as bovine growth hormone, herbicide-tolerant crop varieties, nitrogen-fixing cereal grains, and so on might affect the structure of agriculture in the future. Kenney et al. have emphasized that agricultural biotechnologies are potentially quite diverse and that how these new technologies will affect farm structures depends heavily on the research priorities that are established.

Molnar and Kinnucan, basing their work on existing theories of and evidence on technological change in agriculture, argue that larger farmers will likely be the first to adopt—and thus be the major beneficiaries of —the new biotechnologies. Kloppenburg has developed a preliminary ex ante assessment of how new biotechnologies applied to corn might affect U.S. farm structure over the coming decades.

Buttel has focused on the first major biotechnology to be applied in agriculture: bovine growth hormone, which can be produced in large quantities by genetically modified bacteria and which increases milk production per cow by up to 25 percent. He has argued that this biotechnology will likely reinforce longstanding structural trends toward a dualistic farm structure. The Molnar and Kinnucan anthology contains several important papers on ex ante assessment of new agricultural biotechnology inputs.

A ninth area of biotechnology-related research has been in providing preliminary assessments of how biotechnology might affect the nonfarm economy of nonmetropolitan areas. Finally, Buttel et al., Kenney and Buttel, and Koppel have done research into the implications of biotechnology for international agricultural research and agrarian transformation in Third World countries.

The sociology of agricultural research, especially its recent focus on biotechnology, has been conducted within the "new sociology of agriculture" tradition to a greater extent than any other area of inquiry, save perhaps for the state and agriculture. In particular, the sociology of agricultural research, although involving some quantitative analysis, has been at the heart of trends in the new sociology of agriculture that have created legitimacy for exploring issues that are largely accessible only through historical-qualitative research methods.

Especially in biotechnology, the impacts of which on farm structure will generally not be manifest for a decade or more, quantitative research based on primary data is virtually impossible, and the quantitative research that has been done often suffers from spurious precision. Given the importance of understanding the likely impacts of these new technologies, the work that is being done is of high priority.

Some rural sociologists, however, will feel less than comfortable with a research tradition in which quantitative methodology and the putative precision it affords are largely absent. Nonetheless, the sociology of agricultural research and biotechnology has been among the most innovative and stimulating areas of recent research in rural sociology and will no doubt have a major impact on rural sociological inquiry in the decades to come.

PATENTS ON LIFE AND SEED MONOPOLIES

Seeds and biodiversity, which have been the common property of farmers and local communities, are being transformed into private property of a handful of corporations—Monsanto, Syngenta, Duport, Dow, Bayer. These chemical giants now control seeds as well as medicine. India, in fact, is losing its biological and genetic resources, the basic wealth of our country, and the biodiversity based knowledge of our local communities mainly through biopiracy.

The piracy and patenting of our rich biodiversity by the MNCs, institutions and individuals especially from the Western world is inflicting great injury to the natural resources of our land and people. If this is not arrested by appropriate legislation and policy measures, there would be incalculable damage done to the nation and the coming generations.

India is a signatory of the two international treaties that have an impact on the nation's biodiversity wealth and on peoples' rights to use it and conserve it responsibly. These are the Convention on Biological Diversity (CBD) and the Trade Related Intellectual Property Rights (TRIPs) Agreement under the WTO. India on behalf of a group of countries including Brazil, Bolivia, Cuba, Dominican Republic, Ecuador, Thailand, Peru and Venezuela has submitted a paper to the TRIPs Council of WTO, on "The Relationship Between the TRIPs agreement and the Convention on Biological Diversity and the Protection of Traditional Knowledge", calling for amendments in TRIPs to harmonise it with the CBD.

However, at the domestic level, India has harmonised its Biodiversity laws to be consistent with the unamended TRIPs agreement. The Government of India has enacted new legislations to implement these international obligations which would far reaching implications for the lives and livelihoods of millions of Indian people. The new corporate-driven intellectual property rights regimes (IPRs), especially the Trade Related Intellectual Property Rights Agreement of WTO is leading to seed monopolies and biopiracy. Seeds have been evolved by nature and farmers over millennia.

This collective, cumulate heritage is now either being destroyed by introduction of monocultures of non-renewable seeds or being hijacked by global corporations through patents and biopiracy. When a seed is patented or covered by breeders' rights, farmers can no longer save or exchange seed freely. Seed saving and seed exchange in defined as a "theft" in intellectual property law. The Indian legislations that have an impact on biodiversity and people's rights are: Patent Act, 2002; Protection of Plant Variety Protection and Farmers' Rights Act, 2001; and the Biological Diversity Act, 2002. The country has amended its Patent Act 1970 for the second time, in May 2002, since TRIPS came into force. The first amendment was to intro- duce exclusive marketing rights and mail box arrangement to implement Art. 70.8 and 70.9 of TRIPS.

There are two amendments in the definition of what is not an invention that has opened the floodgates of patenting of genetically engineered seed. First, in Section 3(i) of the Patent Act, 1997, the word "plants" have been omitted. According to Section 3(i), the following is not an invention: 'Any process for the medical, surgical, creative, prophylactic or other treatment of human beings or any process for a similar treatment of animals or plants or render them free of disease or to increase their economic value or that of their products.'

The omission of "plants" from this section implies that a method or process modification of a plant can now be counted as an invention and can hence be patented. Thus the method of producing Bt. cotton by introducing genes of a bacterium Bacillus thurengensis in cotton to pro- duce toxins to kill the bollworm can now be covered by the exclusive rights associated with patents. In other words, Monsanto can now have Bt. cotton patents in India.

The second amendment has also added a new section 3(j). This allows production or propagation of genetically engineered plants to be counted as an invention, and hence patentable. The section 3(j) excludes as inventions "plants and animals...including seeds, varieties and species and essentially biological processes for production or propagation of plants and animals". However, the emergence of new biotechnologies is often used to define production of plants and animals through genetic engineering as not being essentially biological.

Without such a clear definition, 3(j) allows patents on GMOs patentability and hence opens the floodgate for patenting transgenic plants. The language of 3(j) is a verbatim translation of Article 27.3 (b) of TRIPs into India law. Article 27.3(b) of TRIPS states: Parties may exclude from patentability plants and animals other than micro- organisms, and essentially biological processes for the production of plants or animals other than non-biological and microbiological processes. However, parties shall provide for the protection of plant varieties either by patents or by an effective suigeneris system or by any combination thereof.

This provision shall be reviewed four years after the entry into force of the Agreement establishing the WTO [10]. In the process the government has undone its own position on W.T.O. in the TRIPS review. Article 27.3 (b) has been under review since 2000. The whole of TRIPs council has to undertake review of TRIPS "in the light of any relevant new developments which might warrant modification or amendment of this Agreement."

The Biodiversity Act passed by the Parliament in December 2002 was to implement India's obligations under the Convention on Biodiversity Conservation (CBD) of the United Nations.

The breakthrough achieved in the CBD was:

1. The need to conserve biodiversity;
2. Recognition of community rights and indigenous culture and knowledge (Art.8 (j));
3. The prevention of adverse impact on biodiversity by mega development projects and by the release of genetically engineered organisms (GMOs), and
4. The potential to prevent biopiracy—the theft and patenting of indigenous knowledge related to biodiversity.

Earlier drafts of the Bill had all four components. However, the draft passed by Parliament has no conservation element, no recognition of community rights to Biodiversity (even though this is necessitated by the Panchayati Raj Amendments of the Constitution and the extension to Schedule Areas) no regulation of adverse impacts.

Given that the Environment Ministry, through the Genetic Engineering Approval Committee has been rushing ahead to commercialise genetically engineered crops like Bt. cotton in spite of known risks to biodiversity and in spite of exaggerated claims of benefits, it is clear that pressures were put on the ministry to drop clauses on regulating adverse impact in earlier drafts. The act is reduced to an act to facilitate access to biodiversity and indigenous knowledge for intellectual property rights. It facilitates and legalises biopiracy, instead of preventing it.

All that biopirates will need is a cosy relationship with the National Biodiversity Authority. In the case of indigenous seeds and agricultural crops, the Biological Diversity Act has allowed full scope for biopiracy. Section (6) on intellectual property has an exemption 6(3):"The provisions of this section shall not apply to any person making an application for any rights under any law relating to protection of plant varieties enacted by Parliament." Laws protecting plant varieties include the Patent (Second Amendment) and the Plant Variety Protection Act. Exemption 6(3) in the Biological Diversity Act in effect says that companies can take varieties farmers have evolved over millennia with unique traits of aroma as in basmati, salinity resistance, drought resistance, flood resistance, and patent the traits and qualities which are a result of farmers breeding. A global movement is calling for a ban on patents on life and recovering of the generic basis of life as "commons" which cannot be owned and privatised.

The African countries have also made this demand in W.T.O. 3. Technologies of dependency and corporate control one of the key areas of globalisation has been the entry of global corporations in the seed sector. Control over seed is control over the food system. Besides con- trolling seed through patents and intellectual property rights, corporations like Monsanto and Cargill (now owned by Monsanto) are using technologies of dependency to force farmers into seed dependency.

The new seeds are non-renewable— they yield a crop, but the crop is useless for seed because of hybrids based on male- sterility, and the impending introduction of "terminator" seeds—seeds which are engineered to be sterile. Corporations are breeding seeds that need chemicals—life herbicide resistant varieties. Even varieties like Bt. Cotton, need more pesticides because they are prone to more diseases and pests.

The Case of Bt. Cotton

The commercial planting of Bt. Cotton in the country took place in 2002, but the cotton crop failed and the government did not clear its cultivation in North Indian states. Leading agricultural scientists, experts and academicians in India have acknowledged the failure of Bt. Cotton on the following counts: drying and falling of squares without boll formation, reduced boll formation, small sized bolls, very short staple length, very little resistance to boll worm, and requiring 2–3 sprays for control of boll worm, not resistant to dry spells, low yields (only 2–3 quintals for MECH 162), low market value and cost-benefit ratio not on par with non-Bt cotton. The then Health Minister Dr. C.P. Thakur, also observed that Bt cotton products could have long term environmental and health effects.

A study by the Research Foundation for Science, Technology and Ecology con- ducted in Maharashtra, Madhya Pradesh, Andhra Pradesh and Karnataka showed that not only did Monsanto's cotton not protect the plants from the American Bollworm, but there was a increase of 250–300 per cent in attacks by non-target pests like jassids, aphids, white fly and thrips. In addition, the Bt plants became prey to fun- gal diseases like root rot disease or fusarium. The Bt. cotton varieties gave very low yields.

Even the staple lengths of whatever little cotton was produced were so short that it fetched a very low price in the cotton market [16]. Corporate seeds are creating a deep crisis for farmers because of high costs of seed, high dependence on costly inputs (pesticides, herbicides) and high levels of unreliability and crop failure. While costs of cultivation are higher because of costly non-renewable seeds and chemicals, the price of agricultural produce is collapsing due to removal of price and import regulation, a direct result of trade liberalisation.

As production becomes more costly, and agricultural prices fall, indebtedness and economic displacement become the fate of the peasantry, played out in its most tragic expression through farmers' suicides. More than 20,000 farmers have committed suicides because of the debt trap of high cost seeds over the last few years.

Farmers are spending ₹ 100,000 per acre and in a good harvest earning ₹ 10,000, a recipe for debt and suicide. Peasant survival, just and remunerative prices, conservation and sustainable use of vital natural resources, including soil, water and biodiversity, healthy food pro- duction, maximisation of nutrition per acre and remunerative farmers' incomes per acre demands that farmers shift from technologies created for corporate control and maximisation of corporate profits to eco-friendly, farmer friendly technologies such as organic farming.

From food first to trade first: How corporations are taking over Indian agriculture through trade Trade liberalisation and globalisation of agriculture is robbing the peasants and landless workers of already low incomes and livelihood security. There are three mechanisms by which the rural producers are either being super-exploited or they are being rendered dispensable.

Destruction of the market support at domestic level both in terms of procurement and in terms of guarantee of Minimum Support Price (MSP). The MSP is to agriculture what minimum wages are in the individual and service sectors, the minimum prices a farmer should receive to cover costs of production and her/his labour. However, as a result of globalisation, the government has started to withdraw from its role in procurement and price regulation. Riots have occurred in different parts of the country with farmers protesting against lack of markets and fall in prices of agricultural commodities. The case of cotton, for instance, has been discussed above.

Diversion from food crops to perishable cash crops and promotion of mono- cultures thus creating market dependency on corporate monopolies. Globalisation policies have promoted the idea that farmers should shift from food grains and stapes to vegetables and fruits. Export promotion zones for fruits and vegetables are a major thrust area in the new agriculture policy. On the one hand this erodes food security for households and the nation. On the other hand it pushes farmers into distress sales, since fruits and vegetables cannot be consumed or stored at the household level.

Removal of Quantitative Restrictions (QRs) on imports and dumping of subsidised, artificially cheap imports. A dispute initiated by the U.S. against India in the WTO forced the removal of QRs. This has translated into destruction of domestic markets and prices in India being perturbed by the artificially low international prices of commodities. While forcing India to remove import restrictions and reduce domestic support to farmers and the poor, the US has further increased its farm subsidies to USD 180 billion over the next six years amounting to USD 20 billion annually. Most of these subsidies go to agribusiness and to capture export markets. In addition, $10 million funding was made available for export promotions by 65 U.S. trade organizations under the 2002 Market Access Programme (MAP) and another $90 million for Market Access under the fiscal year 2002.

The Farm Bill 2002 also provides that MAP funding be increased to $200 million by 2006. Transnational agribusiness giants like Cargill, ADM and Conagra are, in fact, the only beneficiaries from the liberalization of imports and removal of import restrictions. They benefit both from using their immense financial clout to depress world prices during procurement and hike it during sales, as well as from the various subsidies that are given to them for both exporting as well as importing, from both exporting and importing countries. A recently released report from the International Agriculture and Trade Policy Institute has shown that in four major U.S. commodities, the level of dumping has increased since 1995 when the W.T.O. came into force, even though the proclaimed aim of W.T.O. is to "reduce distortions in trade".

From 1995–2001 dumping jumped from 23–44 per cent in the case of wheat, 9–29 per cent in the case of soya beans, 11–33 per cent in the case of maize, from 17–57 per cent in the case of cotton [6]. According to the World Bank, low cotton prices in U.S. resulting from high subsidies are costing African countries $250 million each year. 4.1. The case of edible oils From 70s to the late 80s, India was a heavy importer of edible oils. In 1986– 1987, India produced 3.9 MT of edible oils, and imported 1.5 MT (28 per cent dependency). However, thanks to the Technology Mission on Oilseeds, the total oilseed production soared from 11.3 MT in 1986–1987 to 21.5 MT in 1993 (10.5 per cent average annual growth).

Imports fell to a negligible 0.35 million tonnes. In 1998, as a temporary shortage of edible oils, combined with unchecked hoarding drove up prices, the government liberalised imports. At the same time the US flooded the world market with soyabean and soya oil, further driving down international prices of soyameal and all edible oils. As a result India has become the largest edible oil importer in the world. 43 per cent of the total edible oil available in the country is imported. As the acreage under oilseeds decreases, with farmers reeling under the price collapse, further imports are envisaged.

Thus, while removing of subsidies and creation of a "level playing field" was the most important argument used by the government for joining WTO in 1974, northern subsidies have actually increased and the playing field has become more uneven. The removal of QRs when combined with dumping becomes a genocidal trade system in which small peasants are wiped out to create global corporate monopolies over food, resulting in the dismantling of the domestic food production system. In the face of rising subsidies and increasing dumping, import restrictions and countervailing duties are a right, a survival necessity, as countries like India, Argentina, Philippines have proposed. WTO has robbed countries of this right through Art. 4.

India and other developing countries should focus on stopping dumping by eliminating Art 4 of Agreement of Agriculture (AoA) which is the basis of the destruction of food security and rural livelihoods in the Third World. Once this crippling clause is removed, countries can start building a global system on citizens initiatives and national priorities that ensures sustainability, supports small farmers, ensures just prices, prevents dumping, protects the countryside and the environment and ensures good, safe, adequate food for all.

Globalisation and the creation of hunger Agribusiness is harvesting both the stocks and agricultural subsidies to gain through agricultural exports. The same companies that gained through deregulated imports are now also gaining through deregulated exports. Millions of Indians go to bed hungry, as the government has dismantled the PDS system, and taken affordable food out of the reach of millions. Foodgrain export is liberalisation's answer to the problem of "over-procurement" from farmers while it will bring in much needed foreign exchange. However, ever since Indian agriculture was subjected to liberalisation, both the share of agriculture exports in India's total exports as well as their value in hard cash has actually declined. The share of agricultural exports in India's total exports declined from 20.33 per cent in 1996–1997 to 18.25 per cent in 1998–1999 and further to 14.04 per cent in 2001–2002.

Foodgrain exports did not benefit the national exchequer each period of large exports has been followed by acute food shortage, and India has had to import the same grain again, often at higher prices. For instance, India exported 2.5 lakh tonnes of wheat during 1987–1988, followed by an import of 18 lakh tonnes in 1988–1989. Foodgrain exports did not benefit the farmer Wheat exports had an inflationary impact on domestic prices, although this did not benefit the farmers, as during the harvesting period the prices are kept low, and even MSP rates are not enforced.

Foodgrain Exports

The push towards exports is neither aimed at helping farmers or the nation's finances; rather, in accordance with the impetus of WTO's Agreement on Agriculture, it is a way of diverting support away from farmers towards traders. While the government dismantles procurement and public food distribution because they are considered subsidies to the people of the country, and are supposed to distort trade, it allows traders to buy bulk grains from its godowns at prices even lower than those offered to the poorest of the poor. Moreover, the annual budgets since liberalization having been adding to the subsidies for the corporate sector–tax holidays for building silos and cold storages, incentives for exporting, subsidized transportation to the ports of the traders' choice. The experience of the 2001 wheat export, for instance, reveals this.

As against an economic cost of Rs 8300 per tonne to the FCI and an open market price of Rs 7000 per tonne, India was offered a price of Rs 4300 per tonne in international market in May 2001. The government allowed FCI to charge any price for exports, as long as it is not below the below poverty line' (BPL) rate. In this scenario, Cargill has emerged as the biggest buyer of subsidised wheat being pushed by India into world markets [17]. The offtake price of wheat for export has also been allowed to far below the BPL price of ₹ 4150 per tonne, to ₹ 3960 per tonne. At the same time, it has increased the above poverty line' (APL) issue prices of wheat to ₹ 610 per quintal, and for rice to ₹ 830 per quintal. This price increase will only ensure that PDS off take will be even lower, so that buffer stocks for exports will be maintained.

Globalisation, as directed by the World Bank/IMF structural adjustment programmes and the WTO has seen a return of famine. The intensification of hunger over the last five years, even while grains are overflowing from the godowns can thus not be attributed to drought and climatic conditions alone, but to the policies that govern people's access to food. The policy changes have consistently attempted to decrease the government's role in ensuring food security and livelihood security for farmers in particular by calling measures meant to aid people as 'trade distorting' and demanding that these be scrapped. The anomaly of the co-existence of burgeoning food stocks and mass starvation is today, as in the late 19th century, the direct consequence of trade-driven agricultural policies.

The 1877 Deccan famine killed over a million people; the trade liberalisation of the last decade of the 1990s has seen India step into the new millennium with over 50 million starving people and millions more with drastically decreasing access to food. An obvious example of the policy shift from people centred concerns to trade- and corporation-centred concerns is the fact that while farmers are not allowed by law to take their produce beyond their state borders, traders can pick them up any- where and take them anywhere. In fact, the government is building super highways, after forcibly taking away land from farmers and communities to connect centres of agricultural production to airports and ports, so that corporations can quickly transport these commodities for export.

The policy changes induced by trade liberalisation include:

- Dismantling the Food Corporation of India (FCI) and reduce its role in procurement from farmers; removal of QRs on imports of food and agricultural products;
- The amendment of the Public Distribution System, to cater only to the BPL category of the population; increase in the central issue price,

from ₹ 450/quintal in April 1995, to ₹ 682 quintal in April 1999, and to ₹ 900/quintal in 2000; and

- Increased price of food available through the targeted PDS to the BPL as a result of transferring 50 per cent of the procurement and distribution costs of the government to this category. These changed policies have had disastrous impacts for both farmers and consumers.

Farmers

Government procurement centres refusing to purchase foodgrains from farmers. The refusal of the government to enforce MSP on private traders and corporations, forcing farmers into distress sale of foodgrains at far below production costs. Dumping of cheap, subsidised agricultural products by other countries in the Indian market. Increasing price of farm inputs, including seeds. Liberalising seed regulations to allow private seed companies sell uncertified seed. Deepening of farmers' debt, increased mortgages and land alienation, increased destitution, suicides and sale of body parts among farmers.

The prosperity that globalisation was suppose to spread is fast proving to be elusive. Trade liberalisation and globalisation has resulted in thousands of farmers sacrificing their lives and livelihoods. In fact, the most prosperous state, also called the breadbasket of India, Punjab, has left behind Andhra Pradesh in the notorious distinction of farmers' suicides.

Consumers

Massive reduction in the number of people accessing food from the targeted PDS due to the inability of the government to identify the BPL category of people. Inability of the majority of even the few BPL identified to purchase food-grain from Fair Price Shops due to increased prices. Drastic increase in food prices—by over 60 per cent since the initiation of trade liberalisation measures, and over 200 per cent in commodities like pulses. Reduced PDS off- take by states due to increased issue prices.

The shift from PDS for all to Targeted PDS was justified on grounds of reducing government expenditure. However, with trade liberalisation, the PDS costs to government have risen from ₹ 5,166 crores in the mid 1990s to ₹ 9300 in 1999– 2000. 3 While the government blames farmers for this increased expenditure, the primary reason is the increased cost of food to consumers as a result of policy changes, resulting in a drastic decline in purchases from the Fair Price Shops.

Declining food production Export-oriented agricultural policies that push the small farmer to destitution on the one hand, and promote cash cropping on the other, have resulted in a steady decline in food production since the early 1990s.

The collapse of domestic support for food production (through dismantling the MSP, rising costs of inputs, crop failure due to uncertified seeds) in the late 1990s has intensified this shift, as farmers are desperate to recover their losses. For instance, there has been a decline in food production to the order of 12.8 per cent in just one year, 2000–2001.

Declining Food Consumption

A major impact of trade liberalisation policies has been a general lowering of food consumption. The per capita cereal consumption has declined from 17 kg per capita per month in the 1950s to 13.5 kg per capita per month in the 1990s. The National Nutrition Monitoring Bureau 1997 data shows a declining trend in consumption in rural India, particularly in cereal and millets, the main source of energy for the poor, from 1990–1995. The National Sample Survey (NSS) rounds starting from the 38th round have documented the decline in cereal consumption 1992, following the implementation of SAPs.

The most important reasons for the decline are: rising food prices, destruction of livelihoods, destruction of the PDS system and shift to export oriented crops. The reduced consumption on the one hand, and decline in production and procurement on the other are directly linked though the food and agricultural policies of the government. Colonialism had destroyed the food sovereignty of the country, forcing changes in cropping from food for local and regional food security to commercial crops. Rice particularly had become a commercial crop even within the country. The emphasis on foreign trade had resulted in large scale famines in the country, forcing Cornelius Walford to comment in 1879 in The Famines of the World—"it is an anomaly that, with her famines on hand, India is able to supply food for other parts of the world".

Following independence, the Government's priority was to ensure that farmers would produce food and thus government procurement (to ensure both that farmers produced food, and got just price for it) and the public distribution system (to ensure that consumers got adequate food at affordable prices) were designed. The need for government involvement in food production and distribution became even more necessary with the Green Revolution that firstly, destroyed regional food security based on diverse cereals and replaced it with just wheat and rice; secondly, concentrated the production of these cereals in just two states—Punjab and Haryana—and thirdly, forced the farmer into the vicious treadmill of costly input (seeds, chemicals water) intensive agriculture.

MOVEMENTS AND SUSTAINABLE DEVELOPMENT

Most of the recent discussion of 'environmental management' in developing countries has paid scant attention to the role of local people. Environmental management is defined almost exclusively terms of a rigidly 'top-down' approach. It is assumed that by incorporating better systems of environmental accounting in the national accounts, and by ensuring that the negative costs of development are allowed for within project budgets, progress will be made towards better management of the environment.

Desirable as it is that economics takes environmental problems seriously, such measures are unlikely to affect the balance of forces which cause poor rural people to place excessive strain on their environments. The 'official' discussion of environmental management in the South makes barely a reference to the increasingly important experiences, and literature, on environmental movements there. The remaining section of this chapter focuses attention upon the need to re-examine our knowledge and understanding of the environment, in the light of the experiences of other cultures far removed, socially and geographically, from the 'core' of the modern food system.

Sustainability and Development

It is clear that rural development projects have often neglected local, ecological and cultural factors in the South. It is also clear that structural factors—which influence people in not acting 'sustainably'—have frequently been ignored. We need, in fact, to take account of both the cultural and structural dimensions in any definition of sustainability. Then we can 'begin to ask whether there are patterns of myth and social organisation, which are more likely to ensure and maintain man's *[sic]* survival base'. It is clear that environmental problems are not new. Historical, and archaeological, evidence suggests that early societies often behaved in ways which were damaging to their environment.

These societies also had to 'solve the basic problem of producing food surpluses and collecting raw materials from rural areas to sustain large urban populations', a problem that faces countries in the South today. However, such societies, unlike the South today, were not faced by twin assaults from rapid economic and demographic growth, on such a scale. They did not exist within a global context which placed their own food systems and values in jeopardy.

If we are to understand fully the relationship between sustainability and development, we must begin by recognizing that they are paradigms with historical as well as geographical antecedents which need to be located within specific social and economic contexts. The point at issue is that societies differ in the way they prescribe courses of action towards the environment. Too often we assume systems possess universalistic characteristics, including the food system.

Such universalistic approaches to social and political systems often leave no room for human agency within the framework of explanation. The question we might address in exploring environmental movements was posed by Giddens in the following way: how does the 'practical consciousness' that people acquire from their society, and from their environment, relate to the 'structural properties of the system'? Environmental movements include movements of resistance to the encompassing food system. Under what circumstances are the structural properties of such a system called into question?

The Epistemological Dimension

To answer this question we might begin, not with an analysis of political conflict over the environment, but with the way in which the 'environment' is understood. Instead of working 'backwards' from social movements to consciousness, we can work 'forwards', from the formation of environmental consciousness to the organization of alternative forms of social action to redress environmental problems.

It is clear that the way in which knowledge is gained about the environment and the way in which this knowledge is used have considerable bearing on the question of human agency, and with it environmental management practices. Howes has argued that in agricultural systems engaged in monocrop cultivation the world view is one in which 'society is conceived as an entity apart from Nature'. It is clear, too, that many non-Western cultures possessed 'scientific theory' before Western science, as we know it, had fully developed. Looking at examples from South Asia, Goonatilake claims that, 'their large-order cognitive maps anticipated the non-manipulable, and distant reality of modern science'.

Furthermore, these kinds of knowledge system often codify knowledge in ways that ensure its cultural survival, long after the reasons for certain observances have passed. In some cases 'after an initial map has been made and legitimized...it tends to continue to be believed for considerable periods of time'. A useful distinction, although it should be emphasized that it exists *within* societies as well as between them, is between what Feyerabend calls *historical traditions* and *abstract traditions.* The two traditions can coexist in a society at any point in time, but some cultures give more emphasis to one tradition than to the other.

Historical traditions involve a different discourse from abstract traditions. It is assumed that 'the objects already have a language of their own' and it is the task of people within a society to learn this language. Their task is to 'learn the language of the objects as they are, and not as they appear, after they have been subjected to standardizing procedures' and experimentation. According to this tradition there is no such thing as pursuing 'objective truth', since truth involves both an object and a subject. Feyerabend compares this with what he terms an abstract tradition.

Abstract traditions are different in that they involve framing statements as a means of building up a model of what happens in nature. According to Feyerabend, 'the statements are subjected to certain rules (of logic, testing and argument) and events affect the statements only in accordance with the rules' It is possible, then, to make scientific statements without personal contact with the objects being described. Both these epistemological traditions exist, at different levels, in most societies, but within the Western scientific tradition much more emphasis is given to abstract traditions.

In this respect the development of Western science is best understood in the context of Judaeo-Christian thinking, which has often appeared inimical to 'scientific' method. For example, Goonatilake argues that religious and philosophical objections to scientific method in Europe would not have occurred in South Asia, where most religious beliefs emphasize the role of both abstract and historical traditions in arriving at valid statements about the world around us. The first point that needs to be emphasized in establishing a framework for understanding environmental conflict in the South is that a society's epistemological framework will influence its understanding of the environment. What we perceive as 'environmental' problems may be interpreted in other cultures as essentially due to the breakdown of social and cultural practices and beliefs.

In addressing these questions we need to be aware that our understanding of Societies very unlike those of the industrialized countries frequently distorts our understanding of their environments. By the same token, the rediscovery that knowledge-systems other than our own have importance for understanding our relationship to the environment has become a central, but contentious, tenet of recent Green thinking. As Luke observes, some Green thinkers have misunderstood the importance of non-Western science. 'Deep ecologists' have sometimes equated small-scale, non-Western societies with 'primal societies', implying that 'primitives' provided an echo of our own pre-Enlightenment past.

This rather naïve view of the superiority of non-Western epistemologies fails to recognize that myths, magic and ritual are 'the functional equivalents (and perhaps the conceptual antecedents) of Enlightenment science and technology'. This kind of simplification and misunderstanding also errs in lifting ecophilosophies out of their cultural context—Taoism or Buddhism—'with little consideration of any cultural grounding', when it is the specific cultural context of environmental epistemologies that gives them their force and vitality. As an approach, the close association of indigenous knowledge with 'answers' to current environmental problems is frequently ethnocentric, a historical and reductionist, substituting descriptions of the world for more rigorous and less ethnocentric analysis.

Other examples exist of societies closer to our own in which environmental problems are constructed from very different political and social milieux. In Eastern Europe today, for example, the commitment to radical political ecology, and the growth of Green movements, is also linked to processes that are not strictly 'environmental' in the narrow sense employed in the West Conflicts over ethnic identity, and the struggles for civil and human rights in contemporary Eastern Europe, have incorporated—and adapted to—environmental demands, and often express these demands in language drawn from the discourse.

The Economic Dimension

The second, important dimension of relations between the environment and society is the economic dimension. The environment is subjected to structural changes over time, linked to economic growth and the development process. If we confine ourselves to the relationship between agriculture and the environment we can identify several features of the development process which have major implications for environmental systems, among them: the production of food as a commodity rather than for household consumption; the adoption of 'high-technology' energy systems on the farm, and in the food processing industries, rather than 'low-technology' energy inputs; and the role of technology in reducing the pressure exercised by population increase on limited natural resources. These factors will be looked at in turn.

Food as an Exchange Value

In some societies agricultural practices underwrite sustainability, in the sense that consideration of the long-term implications of current agricultural practices is enshrined in farm management practices. For example, Wilken shows how some small farmers in Central America, without recourse to fossil fuels, effectively 'bank' their own labour, and that of their household, investing labour in resource management practices that make sense within a longer time-frame. They are seeking to reduce future operational costs on the farm by making major current investments in labour-intensive management now.

As we have argued above, in some societies such practices become ritually encoded. Rappaport shows how, under stable conditions, the ritual enactment of myths may not only help to maintain the social cohesion of the group (a point often made by anthropologists) but also help to sustain the natural environment. Of course, the value placed on a resource is linked to other social referents, such as the concept of ownership, as well as economic factors like the cost of extraction.

The development of wider markets for food products, and the adoption of production techniques that depend on sophisticated off-farm processing and industrial inputs, does not necessarily mark the point at which food becomes a commodity—in most societies food has been a commodity for a long time. However, it does mark the point at which the conversion of food from a use value to an exchange value has the greatest impact on the way the environment is managed, and ultimately socially constructed.

Food is taken from the province of reproduction in small-scale societies and transferred to that of production and consumption in more complex societies. Eventually, in what we can term 'post-industrial societies', the locus of dissent shifts again, this time from production relations to those of consumption. Some of the most important implications for the evolution of the food system are represented by resistance to the introduction of new, processed 'convenience' foods.

Food and Energy Transfers

The food system also affords an excellent example of the way that energy use is linked to the capacity of ecosystems to retain, as well as transfer, value. The process of development, as we have seen, is usually considered in relation to the movement of capital and labour, rather than the exploitation of natural resources. The structural and spatial aspects of transfers of value from one part of the (developing) world to another part of the developed world have been emphasized by O'Connor and Smith. Without dissenting from the significance of these forms of surplus extraction, it is important to emphasize that transfers of energy from within one system to another can also be seen as transfers of value.

Bunker uses both ecological principles, and the laws of thermodynamics, to assert that 'our calculus of value must include not just the labour and capital incorporated into commodities, but all forms of energy and matter'. In his view, 'the embodiment of energy in economic and social organization encompasses far more of the essential differences, and relations, between core and periphery, than measures linked to commodity production and exchange can'.

Population Increase and Carrying Capacity

Economics has also experienced considerable difficulty recently in incorporating the full implications of changes in environmental systems within its analytical framework. The Malthusian tradition emphasized that population pressure placed limits on the capacity of natural resources to feed the population. In fact, like many economists, Malthus did not have a great deal to say about the environment. Barbier reminds us that the pessimism of the Malthusian tradition was only matched by the optimism of the Ricardian view of scarcity. This placed emphasis on 'relative scarcity' rather than 'absolute scarcity'; as resources became used up substitutes were found for them, extending the time-frame in which the environment could facilitate economic growth.

The problem of carrying capacity was not solved but it was 'put back', and technology was the major factor in effecting this change. Later neo-classical thinking, represented by Alfred Marshall, took an even more optimistic position than Ricardo; no tendency was predicted for profits, wages and rent to decline with increased scarcity of natural resources. The implication was that we were entering an 'age of substitution', the environment could be exploited without degradation being inevitable, indeed the scarcity of non-renewable resources might attract a higher price, and therefore lead to their better conservation. This view of 'substitution' still plays a large part in the way we conceptualize the environment today, and has been given added impetus by developments in biotechnology.

The importance of the economic dimension is that it necessarily involves considering the conversion of natural capital into social capital, economic processes of growth and development do not simply modify, they actually transform the environment. In the course of these transformations the development process 'hands back' to nature some serious problems, including that of sustaining economic levels of production and consumption.

The Political Dimension

The development of the modern food system has brought new urgency to another important dimension: the political dimension. The wider parameters of the food system have introduced this dimension into social relations on the farm, relations which were previously mediated by more endogenous factors. Linkages between agricultural producers, suppliers and markets have necessitated more joint decision making, and have led farmers and the food industry into political alliances. As producers and consumers have become incorporated into wider market systems, their response to their tacit incorporation has often taken the form of resistance, in which class, gender or ethnic subordination is contested.

The 'fields of resistance' that result have often served as rallying points for disaffected groups with different political agendas. The concept of fields of resistance we take from Foucault. It is used to distinguish between different responses to the 'individualising techniques of power', through which systems are able to exert social control and often develop a political consensus. The individual's response to the management of this consensus lies at the heart of the resistance to industrially produced food in the developed countries. Attempts to create a consensus around modern agriculture and food policy have recently broken down.

Among the forms of resistance which environmental movements express we can include:

1. *Exploitation,* through which the individual is controlled via market forces and the production process;
2. *Domination,* through which the individual's identity is defined in terms of gender or ethnic roles; and
3. *Subjection,* where the controlling effect of dominant belief systems in the society tie the individual to an ideological position.

An analytical framework based on fields of resistance is of use in exploring the political changes that accompany the development of the modern food system, and the transformation of natural resources. For example, in the case of farmers in the United Kingdom the growth of market forces and production relations on the farm proved to be the decisive factor, at the turn of this century, in bringing about national organizations of both farmers and farm workers from different class positions. However today, farmers and landowners are united in their resistance to restructuring brought about by the market, as well as to government policies on conservation or 'countryside management'.

Farmers have become dependent on the state, and at the same time resentful of their dependence—an ambiguity behind the attraction of agrarian ideology, especially in countries like France and Germany, where farms were smaller and the dislike of state corporatism more pervasive. By contrast, the importance of *subjection,* as a field of resistance, has grown with the increasing effect of modern agricultural practices on the environment. The demand for more 'organic' food, the interest in preserving the physical condition of the countryside, and political mobilization around issues like animal rights and food quality have brought a new urgency to the interface between agriculture and the environment. The political dimension exists in any set of environment-society relations. However, we need to ask why resistance sometimes finds expression through collective activity (land-rights, urban struggles, etc.) and why it sometimes implies withdrawing from the wider public policy sphere into an 'alternative' community or individual 'survivalist' position, at odds with the mainstream political system.

These questions inevitably raise wider issues concerning the conflicting uses to which the environment can be put. The conflicts that surround resistance to 'development' bring into the open alternative strategies for utilizing local knowledge as well as conserving nature. The analysis, at this point, cannot be confined to the food system, and embraces the trajectory under which autochthonous alternatives to development diverge from mainstream economic forces under the direction of capital located in the North. These wider issues are discussed at length in Redclift.

Having considered the principal features of the modern food system from a historical and cross-disciplinary perspective, it is necessary to take the analysis forward into the realm of contemporary culture and politics. This involves analysing the growth of opposition to changes in the food we consume. This chapter has been principally concerned with exploring the effect of the developing food system on rural land use. Increasingly, the focus of concern, for interest groups, politicians and the consumer, has shifted to what we eat, and changes in our diet in turn reflect wider changes in consumption and the use of leisure. The food system assumes its most radical and yet most familiar form in the supermarkets where we in the developed countries buy most of our food. Resistance to changes in food marketing, processing and retailing have assumed a greater importance in recent years. Today they represent a veritable counter revolution.

4

Virus

IMPACT OF VIRUS DISEASE ON SQUASH

Squash, like the other major cucurbit crops, is susceptible to infection by several viruses. The most damaging are cucumber mosaic virus (CMV), papaya ringspot virus watermelon strain (PRSV-W), watermelon mosaic virus 2 (WMV2) and zucchini yellow mosaic virus (ZYMV). While the extent to which viruses cause economic damage in cucurbits, and squash in particular, is not well documented, the problem is extensive enough that diligent effort to breed squash resistant to CMV, PRSV-W, WMV2 and ZYMV has been a goal of breeders for many decades.

Disease breeding in squash

Sources of genes for resistance or tolerance to viruses have been reported in various relatives of squash and in Cucurbita pepo itself. However, introducing useful levels of resistance to more than one virus into commercially acceptable cultivars has been difficult to achieve. While releases of zucchini-type and English marrow-type varieties carrying resistance to a single virus (CMV) has been achieved, breeding for multiple virus resistance has only recently been successful.

A summer squash cultivar, Whitaker, has recently been developed by researchers at Cornell University. This cultivar is resistant to multiple pathogens, including viruses and fungi. The difficulty of this accomplishment is emphasized by the fact that Whitaker was the result of work conducted by a team of researchers for more than a decade.

Because the sources of resistance were related species not normally sexually compatible with C. pepo, these researchers had to employ a commonly used breeding tool, embryo rescue, to overcome the barriers to crossing between species.

SQUASH AS A SPECIES

Cucurbita pepo is a crop native to North America. It consists of several subspecies and varieties that are completely interfertile, and are therefore presently considered part of a single species. C. pepo ssp. pepo includes cultivated pumpkins, marrows and a few ornamental gourds; C. pepo ssp. ovifera var. ovifera includes cultivated crookneck, scallop and acorn squashes, and most ornamental gourd cultivars; C. pepo ssp. ovifera var. texana includes free-living populations found in Texas; C. pepo ssp. ovifera var. ozarkana includes free-living populations found in the Mississippi Valley and the Ozark Plateau; C. pepo ssp. fraterna includes non-cultivated populations found in northeastern Mexico.

Populations of C. pepo exist today under non-cultivated conditions in the United States. These free-living populations range from northeastern Mexico and Texas, to Alabama, and through the Mississippi Valley to Illinois. Some free-living populations, especially in Arkansas, Louisiana and Mississippi, occur as weeds in soybean and cotton fields, and are difficult to control.

Gene exchange between free-living and cultivated populations has been documented. Furthermore, some free-living populations may have originated as escapes from cultivation, which might have undergone subsequent introgression with other nearby cultivated, or other free-living populations.

Genetic engineering of squash

Because of the need to obtain useful virus resistance in a commercially acceptable cultivar, an alternative to breeding was attempted: the direct introduction of genes conferring resistance into already known commercial genotypes. The genetic engineering approach was to introduce the coat protein genes of the viruses in question, a strategy first successful in tobacco, and subsequently shown to be effective in cucurbit species.

Constructs containing viral coat protein genes were introduced into proprietary inbred lines of C. pepo ssp. ovifera var. ovifera via Agrobacterium-mediated transformation. The engineering of the constructs used in the transformations, as well as the transformation procedure, are described in Tricoli et al.. Based on greenhouse testing and small-scale field tests, two lead lines, ZW20 and CZW3, were chosen as parents of commercial hybrids. The characteristics of these lines are described in detail by Tricoli et al.

ZW20

Line ZW20 was transformed with disarmed Agrobacterium harboring a binary Ti-plasmid. This plasmid contained expression cassettes consisting of the coat protein genes of zucchini yellow mosaic virus (ZYMV) and watermelon mosaic virus 2 (WMV2) under the control of the cauliflower mosaic virus 35S promoter. The T-DNA also contained, as a selectable marker, a neomycin phosphotransferase II (NPTII) gene from E. coli transposon Tn5, under the control of the cauliflower mosaic virus 35S promoter.

The transformation event contained multiple insertion events, some of which were structurally complex. Among these insertion events were two loci containing genes encoding both the ZYMV and WMV2 coat proteins, but not the NPTII gene. One locus provided a moderate level of resistance to the two target viruses, while the other provided a high level of resistance.

Advanced breeding lines were made homozygous for the high-resistance locus, but were segregating for the moderate-resistance locus. During this process, the NPTII marker was also eliminated by selection of lines that not only were homozygous for the high-resistance locus, but had also lost the selectable marker through segregation.

CZW3

Line CZW3 was transformed with the plasmid pPRBN-CMV73/ZYMV72/WMBN22, containing expression cassettes for the coat protein genes for ZYMV, WMV2 and cucumber mosaic virus (CMV), as well as the NPTII selectable marker. Line CZW3 was a simpler transformant, consisting of only one insertion event, containing all three coat protein genes, as well as the selectable marker. Advanced breeding lines were homozygous for the transformed locus.

Field Results

Field trials at a number of locations showed that the lines ZW20 and CZW3 were effective in controlling infections by ZYMV, WMV2, and - in the case of CZW3 - CMV as well. Tricoli et al. mechanically inoculated two different ZW20 sublines, including one homozygous for the high-resistance locus. This subline showed only 20% infection after inoculation with WMV2, and no infection at all after ZYMV infection, compared with 100% infection of non-transgenic controls.

The same researchers studied the resistance of CZW3 after mechanical inoculation with CMV, ZYMV and WMV2 individually, and in a mixed infection. Segregating inbred lines were tested, as well as a hybrid in which the segregating inbred was used as a parent. After individual and mixed inoculation of transgenic segregants, only one plant out of a total of 146 showed mild symptoms, while all of the 103 non-transgenic segregants as well as all of the 259 non-transgenic control plants became severely symptomatic.

Further tests were carried out in the field, comparing virus resistance in these lines after mechanical and aphid inoculation. Other tests studied the economic benefit of the resistance provided by lines ZW20 and CZW3 either as an inbred line or as the parent to a hybrid. These tests confirmed that virus resistance conferred by the coat protein genes was able to dramatically increase crop yield and economic return under the pressure of naturally vectored disease in the field.

USDA EXEMPTION

The determination of efficacy against viral disease under field conditions led to the preparation of a petition for a determination of non-regulated status from the United States Department of Agriculture, Animal and Plant Health Inspection Service (USDA-APHIS). USDA-APHIS is the primary agency responsible for regulating work, in the greenhouse and field of transgenic plants. It also has the authority to grant non-regulated status to transgenic plants that in its opinion no longer present a plant pest risk.

The main responsibility for conducting an assessment of safety rests with USDA-APHIS. The assessment relies upon a review of the information package supplied to the agency by the petitioner, but the agency can and does seek out information on its own, and relies on the expertise of the reviewers on its staff. All relevant information is used by the agency to conduct an environmental assessment, and - if the assessment allows - grant non-regulated status.

One guiding principle followed by the USDA-APHIS in assessing risk is to compare the characteristics of the transgenic plant with the characteristics of plants that are produced by traditional plant breeding. Plants produced by traditional breeding are not regulated, because history provides sufficient assurance of safety.

Thus, if the transgenic plant were determined to pose a risk different from that posed by plants already existing in nature or plants that would be produced by traditional breeding, then non-regulated status would be withheld. If, on the other hand, the risks presented by the transgenic plant are not different from those that are presented by traditionally bred varieties, then non-regulated status can be granted.

Since USDA-APHIS relies heavily (but not exclusively) on the information contained in the petition for non-regulated status, information in such a petition should include the following:

- A description of the biology of the plant prior to the transformation. The biological description should include detailed taxonomic information.
- Relevant experimental data and publications relating to the work conducted on the transgenic plant.
- A description of the differences in genotype between the transgenic plant and the corresponding non-transgenic plant. Information must be provided on the organisms from which the transgenes or other control elements were obtained, the method of transformation and the vectors used during the transformation process, as well as the inserted genetic material and its product(s). USDA-APHIS, the Canadian Food Inspection Agency, and Health Canada, have standardized the types of molecular and expression data submitted in a petition for non-regulated status and the criteria by which they should be judged.
- A description of the phenotype of the transgenic plant, including

any known and potential differences from the corresponding non-transgenic plant, in order to substantiate the claim that the transgenic plant is unlikely to pose a greater plant pest risk than the non-transgenic plant from which it was derived.

This information should include the following: plant pest risk characteristics, disease and pest susceptibilities, expression of the gene product, new enzymes, or changes to plant metabolism, weediness of the transgenic plant, impact on the weediness of any other plant which is sexually compatible with the transgenic plant, agricultural or cultivation practices, effects of the transgenic plant on non-target organisms, indirect plant pest effects on other agricultural products, transfer of genetic information to organisms with which it cannot interbreed, and any other information which might be relevant. Any information indicating that a transgenic plant may pose a greater plant pest risk than the corresponding non-transgenic plant must be included.

Field test reports for all trials of the transgenic line in question that were conducted under USDA-APHIS permit or notification. The field test reports should include any observations of deleterious effects on the environment caused by the transgenic plant. Based on information received by USDA-APHIS from various sources including the petition, the Agency granted non-regulated status to ZW20 and CZW3, and published their support for this decision.

In keeping with their goal of considering all information that might be available to them, USDA-APHIS made use not only of data supplied by the petitioners, but also relied on the literature, external expert opinion, and comparison with traits that were already familiar to agriculture. This approach acknowledged that scientifically valid risk assessment could be conducted by relying on information obtained from a wide variety of sources.

Weediness

Of particular concern in determining the plant pest risk of transgenic squash was the potential for the crop itself to become a weed, or for new weeds to arise as a consequence of gene flow from the crop to sexually compatible wild populations. This question was especially important because the transgenic squash would be released in a region that was known to be the centre of origin for C. pepo.

After evaluating all the information available, USDA-APHIS concluded that ZW20 and CZW3 were unlikely to cause a weed problem or a threat to genetic diversity of free-living squash if released from regulation. With regard to causing a weed problem, the agency concluded that the transgenic lines were unlikely to become weeds themselves, and that the virus-resistance traits were unlikely to cause free-living squash to become weeds. The conclusion that the transgenic lines themselves were unlikely to become weeds was based on three reasons. First, the inbred lines used in the transformations belonged to the yellow crookneck squash type (C. pepo ssp. ovifera var. ovifera), which did not exhibit weedy characteristics and was not considered to be a weed in standard texts and weed lists.

Second, introduction of CMV (for CZW3) as well as ZYMV and WMV2 resistance into the crop was not different from other disease-or pest-resistance traits already present in the C. pepo gene pool or introduced into it by traditional breeding. These traits had already been introduced into squash, and even into some commercial cultivars, yet these cultivars had not become weeds.

Finally, the two transgenic lines did not exhibit any other characteristics that were more 'weedy' than those exhibited by traditionally bred squash cultivars. Aside from considering the potential problem caused by the transgenic lines themselves, USDA-APHIS also evaluated the possibility that traits transferred from ZW20 and CZW3 to free-living relatives would cause those free-living plants to become more 'weedy'. USDA-APHIS concluded that this problem was unlikely to happen.

The agency assumed that the transgenic lines and their cultivated progeny would be able to pollinate free-living squash plants and transmit the virus resistance traits to them. This undoubtedly had occurred in the past with other disease- or pest-resistance traits that had been introduced into C. pepo. However, despite this opportunity, no weedy hybrid progeny had emerged.

Since there was no evidence to suggest that the transgenic traits would behave differently from the traditionally bred traits, the historical lack of weed problems provided a sufficient level of assurance that crosses between transgenic and free-living squash were unlikely to create a weed problem as well. Furthermore, the cultivated squash genotypes that were used for transformation possessed characteristics that placed those plants at a selective disadvantage in a non-cultivated habitat.

The transmission of these traits to free-living relatives along with the transgene would place recipient plants at a disadvantage, further reducing their weedy potential. Finally, surveys of free-living squash populations taken by the applicant failed to detect infections of CMV, WMV2 or ZYMV. Therefore, there was no evidence that viruses limited the populations of free-living squash.

Consequently, it was not likely that virus resistance would release these populations from control, nor was it likely that virus resistance conferred a significant selective advantage in free-living habitats. The lack of selective pressure exerted by viruses was also cited by USDA-APHIS as the reason for concluding that genetic diversity in free-living populations would not be reduced. Without selection pressure, there was no reason to expect selection of a subset of genotypes into which the virus resistance would have been crossed.

TRANSCAPSIDATION/RECOMBINATION

The commercial introduction of viral sequences via transgenic plants - and in particular viral genes encoding viral coat proteins - present the potential for interactions between the transgenes or their products, and the viruses to which the transgenic squash might remain susceptible.

These interactions include transencapsidation of heterologous viral RNA by coat proteins expressed in the plants, phenotypic mixing between viral coat protein subunits expressed by the transgenic plant or by the transgenic plant and infecting viruses, and recombination between transgenic viral components and other viruses infecting the transgenic plants.

DeZoeten pointed out that these phenomena have the potential to produce new viruses. USDA-APHIS concluded that the risk of transencapsidation and phenotypic mixing in transgenic virus-resistant plants would be no greater than their likelihood of occurrence in non-transgenic plants already in agriculture. Data submitted by the petitioners, and other information gathered by the agency, showed that the amounts of viral coat protein produced in transgenic plants resistant to CMV, ZYMV and WMV2 were less than the amounts found in susceptible, virus-infected plants.

With respect to this set of data, USDA-APHIS noted that ZYMV and WMV2 coat protein levels increase when ZW20 plants were inoculated with ZYMV and WMV2 singly or combined. The agency speculated that this increase might be due to limited replication of the viruses in inoculated ZW20 plants, or because the coat proteins produced by the transgenes were stabilized by limited replication of the inoculated viruses.

VIRUS

A virus is a small infectious agent that can replicate only inside the living cells of organisms. Most viruses are too small to be seen directly with a light microscope. Viruses infect all types of organisms, from animals and plants to bacteria and archaea. Since the initial discovery of the tobacco mosaic virus by Martinus Beijerinck in 1898, about 5,000 viruses, although there are millions of different types. Viruses are found in almost every ecosystem on Earth and are the most abundant type of biological entity.

The study of viruses is known as virology, a sub-speciality of microbiology. Virus particles (known as virions) consist of two or three parts: the genetic material made from either DNA or RNA, long molecules that carry genetic information; a protein coat that protects these genes; and in some cases an envelope of lipids that surrounds the protein coat when they are outside a cell. The shapes of viruses range from simple helical and icosahedral forms to more complex structures. The average virus is about one one-hundredth the size of the average bacterium. The origins of viruses in the evolutionary history of life are unclear: some may have evolved from plasmids–pieces of DNA that can move between cells–while others may have evolved from bacteria. In evolution, viruses are an important means of horizontal gene transfer, which increases genetic diversity. Viruses spread in many ways; plant viruses are often transmitted from plant to plant by insects that feed on sap, such as aphids, while animal viruses can be carried by blood-sucking insects. These disease-bearing organisms are known as vectors. Influenza viruses are spread by coughing and sneezing.

The norovirus and rotavirus, common causes of viral gastroenteritis, are transmitted by the faecal-oral route and are passed from person to person by contact, entering the body in food or water. HIV is one of several viruses transmitted through sexual contact and by exposure to infected blood. Viruses can infect only a limited range of host cells called the "host range". This can be narrow or, as when a virus is capable of infecting many species, broad. Viral infections in animals provoke an immune response that usually eliminates the infecting virus. Immune responses can also be produced by vaccines, which confer an artificially acquired immunity to the specific viral infection. However, some viruses including those causing AIDS and viral hepatitis evade these immune responses and result in chronic infections. Antibiotics have no effect on viruses, but several antiviral drugs have been developed.

HISTORY OF VIRUS

Louis Pasteur was unable to find a causative agent for rabies and speculated about a pathogen too small to be detected using a microscope. In 1884, the French microbiologist Charles Chamberland invented a filter (known today as the Chamberland filter or Chamberland-Pasteur filter) with pores smaller than bacteria. Thus, he could pass a solution containing bacteria through the filter and completely remove them from the solution. In 1892, the Russian biologist Dmitry Ivanovsky used this filter to study what is now known as the tobacco mosaic virus.

His experiments showed that crushed leaf extracts from infected tobacco plants remain infectious after filtration. Ivanovsky suggested the infection might be caused by a toxin produced by bacteria, but did not pursue the idea.

At the time it was thought that all infectious agents could be retained by filters and grown on a nutrient medium–this was part of the germ theory of disease. In 1898, the Dutch microbiologist Martinus Beijerinck repeated the experiments and became convinced that the filtered solution contained a new form of infectious agent. He observed that the agent multiplied only in cells that were dividing, but as his experiments did not show that it was made of particles, he called it a contagium vivum fluidum (soluble living germ) and re-introduced the word virus.

Beijerinck maintained that viruses were liquid in nature, a theory later discredited by Wendell Stanley, who proved they were particulate. In the same year Friedrich Loeffler and Frosch passed the first animal virus – agent of foot-and-mouth disease (aphthovirus)–through a similar filter. In the early

20th century, the English bacteriologist Frederick Twort discovered a group of viruses that infect bacteria, now called bacteriophages (or commonly phages), and the French-Canadian microbiologist Félix d'Herelle described viruses that, when added to bacteria on agar, would produce areas of dead bacteria.

He accurately diluted a suspension of these viruses and discovered that the highest dilutions (lowest virus concentrations), rather than killing all the bacteria, formed discrete areas of dead organisms. Counting these areas and multiplying by the dilution factor allowed him to calculate the number of viruses in the original suspension. Phages were heralded as a potential treatment for diseases such as typhoid and cholera, but their promise was forgotten with the development of penicillin.

The study of phages provided insights into the switching on and off of genes, and a useful mechanism for introducing foreign genes into bacteria. By the end of the 19th century, viruses were defined in terms of their infectivity, their ability to be filtered, and their requirement for living hosts. Viruses had been grown only in plants and animals. In 1906, Ross Granville Harrison invented a method for growing tissue in lymph, and, in 1913, E. Steinhardt, C. Israeli, and R. A. Lambert used this method to grow vaccinia virus in fragments of guinea pig corneal tissue. In 1928, H. B. Maitland and M. C. Maitland grew vaccinia virus in suspensions of minced hens' kidneys. Their method was not widely adopted until the 1950s, when poliovirus was grown on a large scale for vaccine production.

Another breakthrough came in 1931, when the American pathologist Ernest William Goodpasture grew influenza and several other viruses in fertilised chickens' eggs. In 1949, John F. Enders, Thomas Weller, and Frederick Robbins grew polio virus in cultured human embryo cells, the first virus to be grown without using solid animal tissue or eggs. This work enabled Jonas Salk to make an effective polio vaccine.

The first images of viruses were obtained upon the invention of electron microscopy in 1931 by the German engineers Ernst Ruska and Max Knoll. In 1935, American biochemist and virologist Wendell Meredith Stanley examined the tobacco mosaic virus and found it was mostly made of protein. A short time later, this virus was separated into protein and RNA parts. The tobacco mosaic virus was the first to be crystallised and its structure could therefore be elucidated in detail. The first X-ray diffraction pictures of the crystallised virus were obtained by Bernal and Fankuchen in 1941. On the basis of her pictures, Rosalind Franklin discovered the full DNA structure of the virus in 1955.

In the same year, Heinz Fraenkel-Conrat and Robley Williams showed that purified tobacco mosaic virus RNA and its coat protein can assemble by themselves to form functional viruses, suggesting that this simple mechanism was probably the means through which viruses were created within their host cells. The second half of the 20th century was the golden age of virus discovery and most of the 2,000 recognised species of animal, plant, and bacterial viruses were discovered during these years. In 1957, equine arterivirus and the cause of Bovine virus diarrhea (a pestivirus) were discovered. In 1963, the hepatitis B virus was discovered by Baruch Blumberg, and in 1965, Howard Temin described the first retrovirus.

Reverse transcriptase, the key enzyme that retroviruses use to translate their RNA into DNA, was first described in 1970, independently by Howard Martin Temin and David Baltimore. In 1983 Luc Montagnier's team at the Pasteur Institute in France, first isolated the retrovirus now called HIV.

CHARACTERIZING AND CLASSIFYING VIRUSES

CLASSIFYING VIRUSES

Virus classification is the process of naming viruses and placing them in a 'family tree' in relation to other viruses. This process is different from the one used to classify plants or animals. Thanks to the fossil record, there are clear connections between most animals or plants. Viruses do not leave such a record, so it is more difficult to determine relationships between them. Another factor causing difficulties in virus classification is their pseudo-living nature, many scientists are debating whether viruses should be considered alive because they are missing several criteria considered important for living creatures. This makes viruses very difficult to place in the current classification system for plants and animals.

Virus classification is currently based on five phenotypic characteristics; morphology, or structure, of the virus; type of nucleic acid, or the genetic material, of the virus; mode of replication; hosts; and the type of disease they cause. There are two classification systems in use today, the Baltimore system and the International Committee on the Taxonomy of Viruses classification guidelines.

The Baltimore classification system was developed by Nobel Prize winning biologist, David Baltimore. This system separates viruses into seven groups, designated by Roman numerals, depending on their type of genetic material, the number of strands of genetic material and their method of replication. There are other classification systems that are based on the morphology of the virus or the disease caused.

These systems are inadequate due to the fact that some diseases are caused by different viruses, the cold or flu are the most common example of this and some viruses look very similar to one another. Another factor is viral structures are difficult to determine under a microscope thanks to their small size. By classifying viruses based on their genetic material, some indication of how to proceed with research is provided because viruses in a category behave in a similar manner. The seven groups are: Group I, double stranded DNA viruses such as the herpes virus and the chickenpox virus; Group II, single stranded DNA viruses such as the parvo virus; Group III, double stranded RNA viruses; Group IV, positive-sense single stranded RNA viruses such as the Sars virus, the yellow fever virus and many other well known viruses; Group V, negative-sense single stranded RNA viruses such as the measles virus, the mumps virus and the rabies virus; Group VI, reverse transcribing RNA viruses such as HIV; and Group VII, reverse transcribing DNA viruses such as the hepatitis B virus. The Group VI viruses use the enzyme to reverse-transcribe their RNA into DNA and then insert the transcribed DNA into the host organism's DNA, where it is replicated whenever a cell divides. The Group VII viruses transcribe their DNA into an RNA form, then transcribe the RNA back into DNA to be inserted into the host's DNA and replicated.

In the early 1990s, the International Committee on Taxonomy of Viruses devised and implemented rules for the naming and classification of viruses. This group still oversees the classification of viruses today.

The ICTV system shares many features with the system used for classifying cellular organisms. There are several differences however; the classification of cellular organisms starts with the Kingdom, while viruses start with Order; another difference is the species name generally takes the form Disease species. For example the classification of the yellow fever virus is; Family - Flaviviridae, Genus - Flavivirus, Species - Yellow Fever virus. Notice this virus does not have an Order classification; that is because recognition of the Orders has been extremely slow, in over ten years only three Orders have been named. Many of the 80 known families remain unplaced. The ICTV are still working on this aspect of virus classification.

Viruses can be defined as "noncellular, submicroscopic, obligatory intracellular parasitescomposed of a proteinaceous covering around central nucleic acid (either DNA or RNA) and capable of self--replication within the living host cells."

Viruses are a unique class of infectious agents truely distinctive by their simple organization and mechanism of replication. They are among the most numerous "microorganisms" on our planet and infect all types of cellular organisms. The usual English plural of "virus" is "viruses" although its Latin plural"vira" may also be used. The study of virus s is called virology and the specialists in the field ofvirology are called virologists.

Viruses exist in two states, extracellular and intracellular. In the extracellular state, a virus is a minute particle containing nucleic acid surrounded by protein and occasionally, depending on the specific virus, containing other macromolecular components. In this typical extracellular form, the virus particle, also called virion, is metabolically inert and does not carry out respiratory or biosynthetic functions. The virion is the structure by which the virus genome is carried from the cell in which it has been produced to another cell where the virus nucleic acid can be introduced.

Once in the new cell, the intracellular state is initiated. In the intracellular state, virus replication occurs; new copies of the viral genome are produced and the components that make up the virus coat are synthesized. Since viruses consist of proteins and nucleic acids, they are actually nucleoproteins (also called nucleocapsid). The protein coat of the virus is called capsid, which is in contact with and encloses the nucleic acid. Capsomersare the individual protein units, which comprise the capsid of viruses. A strongly pathogenic virus is calledvirulent.

Vector indicates an organism, usually an insect, that carries and/or transmits disease causing viruses. Aster yellow virus was the first plant virus shown to multiply in insect vectors (i.e., propagative virus) Satellite tobacco necrosis virus (STNV) is the smallest known plant virus.

Pox virus is the largest known animal virus. First successful vaccination against small pox was carried out by E. Jenner. First successful isolation of a virus, the bacteriophage-WLL from the bacterium E. coli, was done by M. Schlesinger (1933).

GENERAL CHARACTERISTICS OF VIRUSES

All viruses are obligate parasites. They multiply only within their living hosts cellsand remain metabolically inert outside the host cell. They are ultramicroscopic and can only be viewed with electron microscope (the smallest known virus is merely 0.002 ?m in diametre, while the largest ones are typically about 0.8?m in diametre). Viruses are, actually, nucleoproteins. The proteinaceous coat (capsid) surrounds the nucleic acid, which forms the central core of the virus particles. The viral genetic material, the nucleic acid, may be either DNA or RNA. The two nucleic acids are never present in a given virus.

These are the nucleic acids of the viruses which areinfectious, and not the protein coat. Viruses are usually so minute that they can easily pass through a filter, which can hold back even the smallest bacteria. Viruses are easily transmitted from infected host to the healthy ones through various agencies. Viruses are so effective that even their smallest amount can cause infection on the host successfully.

Viruses can easily be crystallized. Viruses are considered to be host-specific and represent obligate parasitism even at genetic level. Since the viruses have no metabolic activities of their own and utilize the metabolism of host cells, antibiotics have no effect on them.

ORIGINS OF VIRUS

Viruses are found wherever there is life and have probably existed since living cells first evolved. The origin of viruses is unclear because they do not form fossils, so molecular techniques have been the most useful means of investigating how they arose. These techniques rely on the availability of ancient viral DNA or RNA, but, unfortunately, most of the viruses that have been preserved and stored in laboratories are less than 90 years old.

There are three main hypotheses that try to explain the origins of viruses:

1. *Regressive hypothesis*: Viruses may have once been small cells that parasitised larger cells. Over time, genes not required by their parasitism were lost. The bacteria rickettsia and chlamydia are living cells that, like viruses, can reproduce only inside host cells. They lend support to this hypothesis, as their dependence on parasitism is likely to have caused the loss of genes that enabled them to survive outside a cell. This is also called the degeneracy hypothesis, or reduction hypothesis.
2. *Cellular origin hypothesis*: Some viruses may have evolved from bits of DNA or RNA that "escaped" from the genes of a larger organism. The escaped DNA could have come from plasmids (pieces of naked DNA that can move between cells) or transposons (molecules of DNA that replicate and move around to different positions within the genes of the cell). Once called "jumping genes", transposons are examples of mobile genetic elements and could be the origin of some viruses. They were discovered in maize by Barbara McClintock in 1950. This is sometimes called the vagrancy hypothesis, or the escape hypothesis.
3. *Coevolution hypothesis*: This is also called the virus-first hypothesis and proposes that viruses may have evolved from complex molecules of protein and nucleic acid at the same time as cells first appeared on earth and would have been dependent on cellular life for billions of years. Viroids are molecules of RNA that are not classified as viruses because they lack a protein coat. However, they have characteristics that are common to several viruses and are often called subviral agents. Viroids are important pathogens of plants. They do not code for proteins but interact with the host cell and use the host machinery for their replication. The hepatitis delta virus of humans has an RNA genome similar to viroids but has a protein coat derived from hepatitis B virus and cannot produce one of its own. It is therefore a defective virus and cannot replicate without the help of hepatitis B virus. Similarly, the virophage 'sputnik' is dependent on mimivirus, which infects the protozoan Acanthamoeba castellanii. These viruses that are dependent on the presence of other virus species in the host cell are called satellites and may represent evolutionary intermediates of viroids and viruses.

Historically, there were problems with all of these hypotheses: The regressive hypothesis did not explain why even the smallest of cellular parasites do not resemble viruses in any way. The escape hypothesis did not explain the complex capsids and other structures on virus particles. And, the virus-first hypothesis contravened the definition of viruses in that they require host cells. However, viruses are now recognised as ancient and to have origins that pre-date the divergence of life into the three domains. This discovery has led modern virologists to reconsider and re-evaluate these three classical hypotheses. The evidence for an ancestral world of RNA cells and computer analysis of viral and host DNA sequences are giving a better understanding of the evolutionary relationships between different viruses and may help identify the ancestors of modern viruses.

To date, such analyses have not proved which of these hypotheses are correct. However, it seems unlikely that all currently known viruses have a common ancestor and viruses have probably arisen numerous times in the past by one or more mechanisms. Prions are infectious protein molecules that do not contain DNA or RNA. They cause an infection in sheep called scrapie and cattle bovine spongiform encephalopathy ("mad cow" disease).

In humans they cause kuru and Creutzfeldt-Jakob disease. They are able to replicate because some proteins can exist in two different shapes and the prion changes the normal shape of a host protein into the prion shape. This starts a chain reaction where each prion protein converts many host proteins into more prions, and these new prions then go on to convert even more protein into prions. Although they are fundamentally different from viruses and viroids, their discovery gives credence to the idea that viruses could have evolved from self-replicating molecules.

MICROBIOLOGY

LIFE PROPERTIES

Opinions differ on whether viruses are a form of life, or organic structures that interact with living organisms. They have been described as "organisms at the edge of life", since they resemble organisms in that they possess genes and evolve by natural selection, and reproduce by creating multiple copies of themselves through self-assembly. Although they have genes, they do not have a cellular structure, which is often seen as the basic unit of life. Viruses do not have their own metabolism, and require a host cell to make new products.

They therefore cannot naturally reproduce outside a host cell - although bacterial species such as rickettsia and chlamydia are considered living organisms despite the same limitation. Accepted forms of life use cell division to reproduce, whereas viruses spontaneously assemble within cells. They differ from autonomous growth of crystals as they inherit genetic mutations while being subject to natural selection. Virus self-assembly within host cells has implications for the study of the origin of life, as it lends further credence to the hypothesis that life could have started as self-assembling organic molecules.

EFFECTS ON THE HOST CELL

The range of structural and biochemical effects that viruses have on the host cell is extensive. These are called cytopathic effects. Most virus infections eventually result in the death of the host cell. The causes of death include cell lysis, alterations to the cell's surface membrane and apoptosis. Often cell death is caused by cessation of its normal activities because of suppression by virus-specific proteins, not all of which are components of the virus particle. Some viruses cause no apparent changes to the infected cell. Cells in which the virus is latent and inactive show few signs of infection and often function normally. This causes persistent infections and the virus is often dormant for many months or years. This is often the case with herpes viruses. Some viruses, such as Epstein-Barr virus, can cause cells to proliferate without causing malignancy, while others, such as papillomaviruses, are established causes of cancer.

HOST RANGE

Viruses are by far the most abundant parasites on earth and they have been found to infect all types of cellular life including animals, plants and bacteria. However, different types of viruses can only infect a limited range of hosts and many are species-specific. Some, such as smallpox virus for example, can only infect one species–in this case humans, and are said to have a narrow host range.

Other viruses, such as rabies virus, can infect different species of mammals and are said to have a broad range. The viruses that infect plants are harmless to animals and most viruses that infect other animals are harmless to humans. The host range of some bacteriophages is limited to a single strain of bacteria and they can be used to trace the source of outbreaks of infections by a method called phage typing.

TRANSGENIC TOBACCO PLANTS RESISTANT TO CUCUMBER MOSAIC VIRUS

Since Nicotiana tabacuum is an amphidipoid species, the tetraplold transgenic plants will segregate extensively in their progenies. It needs to be propagated several generations before obtaining a homozygous line. In order to speed up this process, a procedure for a fast production of homozygosity of transgenic plants was devised in our work. Leaf discs of diploid tobacco plant derived from anther cultures of a commercial cultivar NC-89 were transformed with a chimeric vector of CP and Sat-RNA genes of CMV.

RESISTANCE TO RICE STRIPE VIRUS IN TRANSGENIC INDICA RICE PLANTS

Rice stripe disease occurs in rice growing areas of China, Japan, Korea and the Commonwealth of Independent State, and causes significant reduction in rice yield. The coat protein (CP) gene of a Chinese isolate of rice stripe virus (RSV-C) was synthesized, cloned and sequenced, which is very similar to the Japanese strain. Cell suspension cultures were initiated from embrogenic calli of rice Annong S-1 (indica rice) by inoculating yellowish, compact and embryogenic calli derived from seeds into a suspension culture medium containing proline and maltose.

After being cultured at 26°C in the dark for about half a year, finely-dispersed and embryogenic suspension culture was established. Before bombardment, the suspension culture was evenly applied onto three-layerfilter-paper discs in a petri dish. Cells were bombarded with 1 ìm diameter tungsten particles coated with DNA of the expression vector pROK2 containing the CP gene under control of the 35S promoter.

The plasmid DNA was absorbed to tungsten particles using a calcium-spermidine precipitation procedure. 2.5 ìL of the coated particles was loaded onto macroprojectile and each dish with the suspension culture was bombarded three times under a partial vacuum. Following bombardment, the suspensions were cultured in a modified N6 medium. Two days later the suspensions were transferred to the same medium containing G418 (40 ìg/mL), which were subcultured every week.

Being subject to G418 selection for two months, white and fast-growing colonies were emerged from the browny cultures. Green plants could be regenerated when the resistant calli were transferred to a differentiation medium. In one experiment ten plantlets regenerated from G418 resistant calli were tested for their transgenic nature by Southern blot analysis using a32P-dCTP-labelled CP gene as a probe.

The genomic DNA extracted from the selected as well as the control plants were digested with EcoR1 and BamH1 and then hybridized with the probe. Two plants showed two hybridization bands of 0.6 kbp and 0.7 kbp corresponding to the CP gene. Western blot and ELISA further analyses demonstrated that CP (32 kDa) was expressed in the transgenic rice plants. Sixteen transgenic thus obtained and 100 non-transformed rice plants were inoculated with RSV-carrying leaf hoppers, 24 days post inoculation feeding, most of the untransformed plants developed chlorotic stripes and chlorosis in the young leaves, the disease index (DI) being 53%.

After 40 days post inoculation feeding, most of the control plants showed severe symptoms, being stunted and the leaves developing chlorotic and brown necrotic streaks, with the DI being 88%. In contrast, the transgenic rice plants subject to the inoculation feeding developed no or mild symptoms. After 24 and 40 days post inoculation feeding, the disease index of the transgenic plants was 15% and 25% respectively.

VIRAL CULTIVATION AND PHYSIOLOGY

Viruses can be cultivated within suitable hosts, such as a living cell. To study bacteriophages, for example, bacteria are grown in a suitable growth medium; then bacteriophages are added. The bacteriophages multiply within the bacteria and increase their numbers substantially. Animal and plant viruses are cultivated in cell cultures.

A cell culture is prepared by encouraging cell growth outside the animal or plant source. The cells are kept alive in a suspension of growth factors within a Petri dish. A thin layer of cells, or monolayer, is then inoculated with viruses, and replication takes place. Fertilized eggs and living animals can also be used to cultivate viruses.

For research study, viruses can be cultivated in large volumes by inoculations to tissue culture systems. After a time, the cells are degenerated, and viruses are harvested. The viral particles are concentrated by precipitation methods and purified by repeated centrifugations. Highly purified viruses can be obtained by crystallization and concentration under established conditions.

VIRAL MEASUREMENTS

Viruses are generally too small to be seen under the light microscope, and an electron microscope is usually necessary to make them visible. Although viruses can be quantified by observation, it is also possible to determine their number in terms of virus infectious units, each of which is the smallest unit that causes a detectable effect when viruses infect a susceptible host. Virus infectious units are expressed per volume of fluid. One method for determining virus infectious units is by the plaque assay.

The plaque assay is performed by cultivating viruses on a "lawn" of host cells and noting the presence of clear areas where viruses have replicated and destroyed the cells. Another way of determining virus infectious units is by cultivating viruses in living animals and determining which dilution of virus is lethal to the animals. The end-point dilution can be calculated by this method.

ANTIVIRAL AGENTS

The antibiotics normally used to treat bacterial disease cannot be used to inactivate viruses because viruses do not perform the biochemical functions that antibiotics interfere with. For example, penicillin is used to interrupt the synthesis of the bacterial cell wall, but viruses have no cell wall. However, there are several nucleotide analog drugs that interfere with viral replication.

Acyclovir, for example, is used against herpes viruses because this drug prevents the synthesis of DNA during viral replication. A drug called azidothymidine (AZT) is used for patients with HIV infection because this drug also prevents the synthesis of DNA. A drug called ganciclovir is used against cytomegaloviruses, and amantadine is useful against influenza viruses. Interferon, a naturally produced antiviral agent approved for certain uses, is a group of proteins produced by host cells after they have been infected by viruses. The interferons do not protect the host cell, but they do provide protection to neighbouring cells against viral replication. Interferons can be produced by genetic engineering methods.

VIRAL VACCINES

Protection against viral disease can be rendered by using a viral vaccine. Viral vaccines can be composed of inactivated or attenuated viruses. Inactivated viruses ("dead viruses") are unable to replicate in host cells because of some chemical or physical treatment.

The Salk vaccine against polio and the yellow fever vaccine are examples. Attenuated viruses ("live viruses") are weakened viruses that replicate at a very slow rate in host cells and generally do not produce any symptoms of disease when inoculated to humans. Attenuated viruses are used in the Sabin polio vaccine and in the vaccines against measles and rubella. The most contemporary vaccines are composed of viral proteins produced by genetic engineering methods. The vaccine for hepatitis B is an example of this type of vaccine.

VIRAL INACTIVATION

Virus particles are composed of nucleic acid, protein, and in some cases, a lipid envelope. As such, the viruses are susceptible to normal inactivation by chemical substances that react with any of these organic compounds. Such things as chlorine, iodine, phenol, detergents, and heavy metals rapidly inactivate viruses. In addition, viruses are destroyed by heating methods used for other microorganisms, and they are very susceptible to the effects of ultraviolet light. Filters can be used to remove viruses from fluids so long as the filter pores are small enough to trap viral particles.

VIROIDS AND PRIONS

PRIONS

Structure

Prions are infectious agents composed exclusively of a single sialoglycoprotein called PrP 27-30. They contain no nucleic acid. PrP 27-30 has a mass of 27,000 - 30,000 daltons and is composed of 145 amino acids with glycosylation at or near amino acids 181 and 197. The carboxy terminus contains a phosphatidylinositol glycolipid whose components are ethanolamine, phosphate, myo-inositol and stearic acid.

This protein polymerizes into rods possessing the ultrastructural and histochemical characteristics of amyloid. Amyloid is a generic term referring to any optically homogenous, waxy, translucent glycoprotein; it is deposited intercellularly and/or intracellularly in many human diseases such as:

- Alzheimer's disease
- Creutzfeldt-Jakob disease
- Down's syndrome

- Fatal familial insomnia
- Gerstmann-Straussler syndrome
- Kuru Leprosy

Replication

The prion is a product of a human gene, termed the PrP gene, found on chromosome 20. This gene contains two exons separated by a single intron. Exon I and Exon II are transcribed and the two RNAs ligated into a single mRNA. This mRNA contains an open reading frame (ORF) or protein coding region which is translated into the PrP protein. The PrP protein is a precursor of the prion protein. It is termed PrP 33-35.

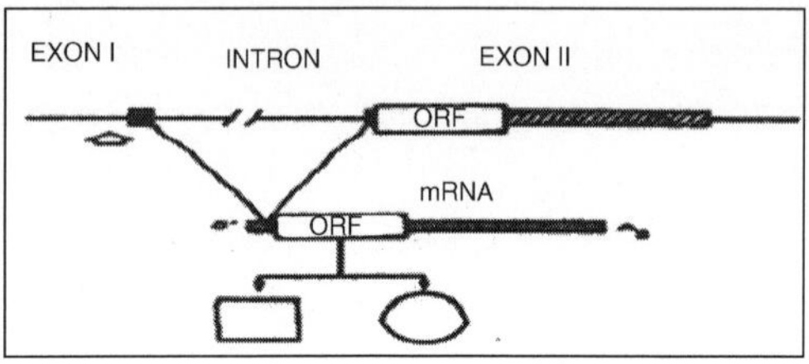

The PrP 33-35 undergoes several post-translational events to become the prion protein (PrP 27-30):

- Glycosylation - at two sites.
- Formation of a disulfide bond between two cysteine residues.
- Removal of the N-terminal signal peptide.
- Removal of the C-terminal hydrophobic segment.
- Addition of a phosphatidylinositol glycolipid at the C-terminal.
- Removal of the N-terminal first 57 amino acids.

In normal cells only the PrP 33-35 protein is synthesized. It is found in the neural cell membrane where it's function is to sequester Cu++ ions. In abnormal ("infected") cells, the PrP 27-30 is produced from the PrP 33-35 protein.

The PrP 27-30 triggers a series of reactions that produce more PrP 27-30 proteins, i.e., PrP 27-30 induces its own synthesis. In addition to the post translational modifications, the PrP 27-30 protein differs from the PrP 33-35 protein in a single amino acid residue. Residue 178 in the PrP 27-30 contains an asparagine residue whereas the PrP 33-35 protein has an aspartate residue at this position. This causes a conformational change in the PrP 27-30 protein from an a-helix to a b-sheet.

This conformational change in the PrP 27-30 protein has three effects:

1. It imparts to the PrP 27-30 protein the ability to induce the same a-helix to b-sheet conformation in the PrP 33-35
   - *Protein*: This is a permanent conformational change. It thus induces its own "replication."
2. The b-sheet-forming peptides aggregate to form amyloid fibrils.

3. The amyloid fibrils kill thalamus neurons through apoptosis, a programmed series of events that leads to cell death.

Pathologies Induced by Prions

All diseases known to be of prion etiology, in animals and humans, are neurodegenerative diseases.

In the human this includes:

- Creutzfeldt-Jakob disease (CJD)
- Fatal Familial Insomnia
- Gerstmann-Straussler syndrome
- Kuru

The pathological and clinical signs of these diseases suggest that they are closely related. In fact they may be variants of the same disorder. All pathological features are confined to the central nervous system. The prion protein accumulates selectively and abnormally in CNS nerve cells during the course of the disease.

PrP 27-30 accumulates within the neuropil where it causes:

- Astrocyte gliosis (an increase in the number of astrocytes).
- Depletion of dendritic spines in neurons.
- Formation of numerous vacuoles in the cerebellar cortex (spongiform encephalopathy).

Amyloidosis - deposition of amyloid in the cerebellar cortex, thalamus, brain stem and in the lumen of blood vessels within the brain. These amyloid plaques consist of discrete eosinophilic glassy-appearing masses, often having radiating amyloid fibrils at their periphery. The plaques are primarily subependymal, subpial and perivascular. Note that the pathology does NOT include any signs of inflammation or fever. This is evidence that the immune system does not respond to the prion protein. Since the prion protein is derived from self this is what you would expect. These pathologies give rise to the clinical symptomology seen in these patients.

These are:

- A long incubation period (several years) which has given rise to the term "slow infection."
- Loss of muscle coordination which leads to a difficulty in walking, indicating a functional disorder of the cerebellum.
- Dementia characterized initially by loss of memory, diminished intellect and poor judgement.
- Progressive insomnia characterized by a marked reduction or loss of the slow-wave and rapid-eye-movement phases.

Transmission

Spread of the disease is via horizontal transmission, i.e., transmission from one person to another, either directly or by fomites or by ingestion of contaminated meat.

Diagnosis

In the past, diagnosis of prion disease was made through examination of brain biopsies taken from patients in advanced stages of the disease or, more commonly, after they had died. In January of 1999 it was found that the prion protein accumulated in the tonsils and could be detected by an immunofluorescence test on tonsilar biopsies. A second test was simultaneously developed which was based on a Western blot. Later that year a third test was developed that had the high sensitivity necessary to detect the prion protein in blood. This test is based on capillary electrophoresis with laser-induced fluorescence. It detects as little as 10^{-18} mole.

VIROIDS

Structure

Viroids are infectious agents composed exclusively of a single piece of circular single stranded RNA which has some double-stranded regions.

Because of their simplified structures both prions and viroids are sometimes called subviral particles. Viroids mainly cause plant diseases but have recently been reported to cause a human disease. Catalytic RNAs are those that have the intrinsic ability to break and form covalent bonds; Viroids are catalytic RNA's (ribozymes) that cleave RNA to produce fragments containing a 5'-hydroxyl and a 2', 3'-cyclic phosphate.

This is a nonhydrolytic reaction in which the same number of phosphodiester bonds are maintained and the transesterification reaction is theoretically reversible. This reaction is considered to play an essential role in the replication of these RNAs *in vivo*. Such reactions are all intramolecular and hence quasi-catalytic with single turnover. These RNAs can be manipulated, however, to provide true catalytic cleavage in trans-reactions.

Replication

Circular, pathogenic RNAs are replicated by a rolling circle mechanism *in vivo*. There are two variations of this rolling circle mechanism:

In the first variation (A), the circular plus strand is copied by viroid RNA-dependent RNA polymerase to form a concatameric minus strand (step 2). Site-specific cleavage (arrows) of this strand produces a monomer that is circularized by a host RNA ligase (step 3) and then copied by the RNA polymerase to produce a concatameric plus strand.

Cleavage of this strand (step 5) produces monomers which, on circularization, produces the progeny circular, plus RNA, the dominant form *in vivo*. In the other variation (B), the concatameric minus strand of step 1 is not cleaved but is copied directly to give a concatameric plus strand (step 3), which is cleared specifically to monomers for ligation to the circular progeny. Those RNAs that self-cleave only in the plus strand *in vitro* are considered to follow this route.

The hepatitis D viroid genome is a minus strand that gives rise to two RNA species. One of these is a mRNA for the delta antigen and the other is a complete complimentary copy (plus strand or anti-genome). The anti-genome acts as a template to make more minus strands. The minus strand self-cleaves and self-ligates. HDV replication takes place in the nucleus but delta antigen is made in the cytoplasm. The delta antigen is the only protein made by the HDV mRNA. It has a +12 charge at physiologic pH, accumulates in the nucleus and binds to minus strand RNA as a dimer. The delta antigen is necessary for viroid assembly but its exact mode of action is unknown.

Human Pathologies Induced by Viroids

The only human disease known to be caused by a viroid is hepatitis D. This disease was previously ascribed to a defective virus called the delta agent. However, it now is known that the delta agent is a viroid enclosed in a hepatitis B virus capsid. For hepatitis D to occur there must be simultaneous infection of a cell with both the hepatitis B virus and the hepatitis D viroid. There is extensive sequence complementarity between the hepatitis D viroid RNA and human liver cell 7S RNA, a small cytoplasmic RNA that is a component of the signal recognition particle, the structure involved in the translocation of secretory and membrane-associated particles. The hepatitis D viroid causes liver cell death via sequestering this 7S RNA and/or cleaving it.

Transmission

The hepatitis D viroid can only enter a human liver cell if it is enclosed in a capsid that contains a binding protein. It obtains this from the hepatitis B virus. The delta agent then enters the blood stream and can be transmitted via blood or serum transfusions.

5

Fungal Endophytes of Plants

INTRODUCTION

The plant is an extraordinarily common source of organic energy. It is thus likely that a huge array of fungi interact with plants. Of these, a range can be isolated from apparently healthy tissue, and many have never been documented to be associated with disease. Others may cause disease when environmental conditions change. Others appear to be the organisms usually associated with initial degradation of the leaf following senescence.

In addition, one specific group of fungi have a close association swith grasses. These fungi are transmitted vertically (through seed) and their impact on the economic value of the host has led to a study of their biology. The grass endophytes, and the related fungi are considered separately from the more common general endophytes.

A wide diversity of fungi are isolated from the tissues of most terrestrial and marine plants, and red and brown algae. Fungi are present in most plant parts, especially the leaves. Where the tissue is apparently healthy, the fungi may be either endophytes, epiphytes or latent pathogens. Endophytes are contained within the plant without disease. Plant tissues remain entire and functional.

ENDOPHYTIC FUNGI

Endophytes colonise plant tissue and remain within the tissue except that fruiting structures may emerge through the surface. Indeed, leaves may be fully colonised by a variety of fungi within a few weeks of emergence. The colonies remain asymptomatic and some in perennial parts may have a very long life.

Endophytic fungi are found in all divisions of fungi so have presumably evolved the association independently on many occasions. The most common endophytes are anamorphic members of the Ascomycota, and they are often closely related to fungi known to cause disease, either in healthy tissue or as secondary invaders of damaged tissue. This suggests that the endophytes may have evolved from pathogens or vice versa. The mechanisms of host recognition and development of colonisation may also be common. A wide range of plants have now been examined for endophytes, and endophytes have been found in nearly all of them.

An enormous number of different fungi can be isolated from plants growing in their native habitat. Most of the fungi are uncommon and narrowly distributed, taxonomically and geographically. However a few fungi are widely distributed with the host, suggesting a long standing, close and mutually beneficial interaction. Indeed, some fungi are found in many different terrestrial hosts, especially endophytes of crop plants. While most information has been gathered from terrestrial ecosystems, fungi are found in algae and seagrasses. Just as we know less about marine ecosystems, our understanding of the biology of marine endophytes is extremely limited and will not be discussed further.

Dispersal of the endophytic fungi remains puzzling. Apart from *Neothyphodium* and related species, endophytes are transmitted horizontally. That is, each plant is colonised by fungal propagules that arrive from the environment. The source of transmission has been determined in only a few cases. Propagules of some endophytes have been found in the body of insect pests of the host. Intriguingly, at least two entomopathogens have been documented as endophytic fungi. Thus insects may disperse the fungi from host to host.

Aerial dispersal either in the wind or on vectors is probably the most common mechanism for fungal dispersal. Endophytic fungi colonise various parts of the plant. Many of the fungi sporulate in culture indicating the potential to release spores in the air. Indeed, sporulation is seen after senescence of plant tissues. However, few cases of dispersal have been documented in the wild and the various mechanisms remain unexplored.

Nutrients are cycled between the host and fungus. The fungus clearly gains a predictable environment in which nutrients are readily available. Thus the benefits are clear for fungi establishing endophytic associations. The loss of plant resources to the fungus, and the potential of some fungi to grow rapidly, indicates that the host regulates development of colonies.

Each plant host has a range of physical, chemical, constitutive and induced controls over the spread of fungi within tissues. An enormous diversity of phenolic and other deterrent plant compounds are associated with the presence of endophytic fungi, in fact more than are associated with potent pathogens in the same host. In addition, presence of endophytes upregulates plant responses to pathogens. In the absence of plant controls, proliferation of endophytes through tissues would be expected.

The reaction of the plant to endophytes suggest that the interaction is one of confinement by the plant. Colonisation by endophytes ranges single cells (*Rhabdocline parkeri*) to patchy distribution through leaves and stems (*Chaetomium globosum*). The presence of specific endophytes, the lack of signs of disease and the long-lived association indicate that further study of the association may yield interesting information of the precise benefits to plants associated with these colonists.

Plants may benefit from the presence of endophytes in many ways. Potential plant benefits have been examined in only a few cases. *Rhabdocline parkeri* produces a compound that reduces needle attack by borers. Metabolites produced by *Phomopsis* sp in cotton appear to deter larvae of *Helicoverpa* from feeding on leaves. The parallels with *Neotyphodium* are clear. In addition, aphids feeding on leaves of cotton may become colonised by *Lecanicillium lecanii,* when conditions permit. Thus the aphid may be killed or it may transfer the fungus to another leaf.

In addition, endophytes may upregulate host responses to pathogens and pests. *Chaetomium globosum* has been shown to increase host resistance to rust and tan spot pathogens in wheat. Direct interactions appear to be too small to measure in this case. Presence of *Lecanicillium lecanii* appears to reduce the feeding by aphids from leaves of cotton. The interaction is probaly due to induction of host responses, which is perplexing because plant regulatory pathways for responses to insects and pathogens are not thought to be complementary. Endophytes appear to have direct and induced effects on plant responses to biotic agents. The interaction with abiotic agents remains largely unexplored.

The broader, ecological function of endophytic associations is still being debated. Many fungi that are associated with the initial stages of litter decomposition are found in healthy tissue of the same plants. Thus they are involved in the initial stages of resource recycling. Endophytes are also associated with aquatic activities. Many aquatic fungi have an endophytic stage in their life cycle. An enormous diversity of endophytes are found. Endophytes are probably associated with a wide variety of host functions.

Latent Pathogens

Latent infection of plants by pathogens is the state where a host is infected by a potential pathogen and the signs of infection are absent. Latent infection eventually results in the expression of macroscopic signs. Disease may be induced by changes to the host physiology due to the activity of the fungus, or changes in the environment that increase the stress on the plant. For instance, a decline in available light, or water, or increased herbivory may trigger development of disease symptoms. In some cases, the signs of disease do not appear until the host tissue has reached maturity or senescence. The causes of disease are possibly due to significant removal of host resources, response to toxins released as secondary metabolites by the fungus, or alteration of host metabolic pathways which reduce host growth rates. At the very least, pathogens would cause dwarfing of the infected host.

The period of latency has been defined as the time between initiation of infection to the expression of macroscopic signs. Thus the length of the latency period may be days or years. Latent infections are important for the damage they cause directly. They also weaken the host plant, predisposing it to infection by other pathogens and pests. Further, infection of one plant is likely to spread to adjacent plants as happens with many pathogens of crops. Because the host lacks obvious signs of the presence of latent pathogens, the presence of latent pathogens is likely to go undetected and their effect ascribed to other causes.

Epiphytes

Fungi that exist on the surfaces of plants are called epiphytes. While some fungi are adapted to the plant surface, the community also includes propagules of air-borne species, fungi that would not otherwise be considered epiphytes. The interaction of epiphytes with their host is largely unknown. The surface of the plant, especially the leaf is a challenging environment. Though plants express metabolites through the cuticle, the surface is dry, waxy and affectedby UV radiation.

Epiphytic fungi differ markedly from endophytes. Epiphytes have many obvious characteristics that enable continuation through the inclement conditions of the leaf surface. In brief, epiphytes are likely to be coloured (melanised), and thus able to resist UV radiation. Some epiphytes can digest lipidic substrates, and thus may utilise the waxy layer covering the leaf. Finally the yeast form has a relatively short life cycle enabling epiphytes to multiply during the short periods of appropriate environment.

Alternaria: A Latent Pathogen of Cotton

Cotton is commonly colonised by Alternaria alternata and A. macrosporum. The fungi are among the first that can be isolated from emergent seedlings. The fungi can be readily isolated from leaf tissue for the rest of the growing season, and spores can be washed from the leaf surface. The leaf tissue is heavily colonised. The fungi rarely cause serious disease. However, when concentration of K in the leaf tissue falls below adequate levels, a purple leaf spot develops which contains Alternaria. Under experimental conditions of K deficiency, inoculation of healthy leaves with Alternaria, results in similar signs. Other factors are also usually associated with development of the disease and causation is apparently multifactorial.

NEOTYPHODIUM AND THE GRASS ENDOPHYTES

The fungal genus Neotyphodium is known for its association with livestock disorders. In particular, N. coenophialum in fescue and N. lolii in ryegrass have been intensely studied. The mechanisms underlying the disorders of herbivores and the other effects that Neotyphodium species have on plants. Many grasses have associations with Balansioid fungi, the most commonly studied associations are between C_3 grasses (cool season growth) and Neotyphodium. C_4 grasses appear to have epibiotic associations more commonly.

Infected Grasses

Many grasses (Poaceae) in the subfamily Pooideae are associated with intercellular endophytic fungi. The fungi are mainly found in leaves, leaf bases, rhizomes and seeds. The association has no external signs and signs of disease are absent. The association between Neotyphodium and C_3 grasses is highly specific. Fungi isolated from one plant species, or even strain, can be inoculated into a second, where it is unlikely to persist. We know little about the mechanism underlying the specificity.

Life Cycle of theFungus

Neotyphodium colonises the intercellular leaf and leaf base tissue. Usually the fungi can be observed as wavy hyphae located immediately below the epidermis. As the flowering stem elongates, the fungus grows with it and infects seeds as they form and develop. The fungus is transferred asexually from plant generation to plant generation within the seed in what is called 'vertical transmission'. The seed germinates and the fungi colonise the elongating leaf tissues. In other words, the fungus is clonal and the plant host sexual. The fungi have not been found outside the association. The fungi have no known sexual system and no other known mechanism to infect seedlings in the wild.

MECHANISMS OF ACTIVITY

Feeding Deterrence

Defence chemicals are produced by the fungus, or the host when infected by the fungus. The chemicals disrupt several metabolic processes in herbivores that use the grasses. The active metabolites are found in at least four broad categories:

The ergopeptides ergovaline and ergovalinine are based on lysergic acid amide. Generally known as ergot alkaloids, they induce various metabolic disorders that reduce reproductive rates and growth of grazing animals. The signs include elevated temperature, vasoconstriction, and reduced prolactin in blood. Serum prolactin is a growth hormone that regulates feed intake in mammals. The ergot alkaloids also influence animal gait. The compounds are found in many other members of Balansioid fungi.

Tremogenic neurotoxins Lolitrem A and B are indole diterpenoids. Along with paxilline alkaloids they induce various stiff legged gaits in sheep grazing on infected grass. They also are toxic to or deter feeding by various insects.

The unusual Pyrrolopyrazine Peramine is thought to influence insect activity. It is widely found to deter insect feeding. While peramine is the only compound so far chemically characterised, related lipophilic compounds have been indicated in some other endophyte associations. Pyrollizidine-based loline alkaloids reduce feeding by some insects, and acts in concert with the other compounds to reduce feeding and reproductive rates of many more insects. They also appear to be a factor in the allelopathic properties of infected fescue.

The chemicals influence livestock, insects, mites,and pathogenic nematodes. Other compounds are believed to influence fungal pathogens. These chemicals serve as feeding deterrents, reduce rates of reproduction and growth. The effect of the chemical is systemic in the plant; while the fungus colonises only the shoot, the deterrents influence shoot and root feeders.

Livestock have reduced rates of growth in cases where endophytic grass is the sole or major source of plant material. The animals tend to develop unsteady gait and stand in water on hot days. The rate of milk production in dairy cattle is also reduced.

Insect attack is reduced in three ways in endophyte infected grasses. Fewer insects land on endophyte infected grasses. Insects will feed for shorter periods and less frequently on infected grasses even when they land and feed. Finally, rates of insect reproduction, specifically number of eggs laid and the viability of eggs, is reduced. Under field conditions, specific populations of fungi and insects have developed, where the toxins produced by the endophyte deplete the insect population, but not completely. The endophyte/grass symbioticum is commonly not adapted to newly arrived insect species. With the arrival of new insects, grazing pressure is quite high until fungal genotypes able to deter the new herbivore become more common in the pasture. Typically, a long established pasture contains a mixture of fungal genotypes.

Drought Tolerance

Endophytic grasses survive under intermittent drought longer than uninfected grasses. The mechanism is unclear, though change in osmotica within the plant tissues following colonisation by the endophyte may be important in some cases. Changes in osmotica suggest that the mechanism is based on changes in hormonal balances, either directly due to the fungi, or indirectly by the fungus influencing plant hormonal balances.

Competition

In natural pastures, Neotyphodium may colonise up to 90% of the grass plants present. Neotyphodium infected plants are more competitive in mixed pastures (eg clover/ryegrass) with uninfected plants. Low levels of infection at sowing (4%) may change to more than 40% within a few years, regardless of whether the pastures are infested with insects, affected by drought or diseased. Established endophyte infected plants have more tillers, and produce more seeds than uninfected plants, regardless of the grazing pressures. The variation between endophyte infected grasses is greater than uninfected grasses and they therefore have a wider region of adaptation than uninfected grasses. We are aware of no examples where the infection resulted in reduced competitive capability. However, in soils where mineral nutrients are readily available, uncolonised plants have better survival in competitive interactions. The competitive interaction appears to favour colonised plants, but the mechanisms underlying the resultant population of plants and fungi are complex and interrelated.

Cost of the Endophyte

Published measures of the metabolic cost of the endophyte suggest that it is undetectable. Indeed, endophyte infection may increase phosphate concentrations in the shoot tissue which influences rates of photosynthesis. Nitrogen use appears to be more efficient in some plant/fungus genotypes. The fungus almost certainly produces the deterrent compounds under direction of the host plant.

The fungus increases phenotypic variation of the plant. The value of measures of metabolic costs are debatable. Measures in lab suggest that the cost is variable and determined in part by the density of the fungus, the presence of a mycorrhizal fungus in the root system and the phosphate nutrition of the plant. In many respects the endophyte/grass association is similar to mycorrhizas, with the added input of chemicals that influence interaction with other organisms.

Applications to Agriculture and Turf Culture

Fungi can be selected for production of specific toxins, and selections can be genetically manipulated in culture using classical techniques. The goal of pasture grass breeders is to produce or select fungi that have only marginal effect on livestock but major deterrent on pests and diseases of the grass. Turf farmers have no interest in grazing by mammals, though they too wish to reduce insect damage. So far, reinoculation of grasses with desirable fungi has proven to be a major stumbling block due to the specificity of the association.

The fungus does not survive in seed in hot conditions. In agronomic systems, the fungi may die out if the seed is stored for longer than 6 months, much less if the temperature of storage exceeds 36°C for more than 6 hours. Many so called "endophyte" seed stocks lack the fungus because storage conditions are primitive and seed is often kept for long periods.

Members of a specific genus of fungus release highly toxic metabolites when in association with the grass host. These metabolites have been studied because of their effect on grazing livestock and herbivorous insect pests. The consequence of endophytes, *sensu stricto,* is that they modify the capacity of the host to function in competitive interactions with other plants. This selective advantage enables increases in the proportion of colonised grasses in pastures and presumably native vegetation. It also places herbivores under selective pressure to evolve resistance to the toxins. The wide array of metabolites found in these fungi is typical of all the Ascomycetous fungi intensively studied, suggesting that formation and excretion of metabolically active compounds is a widespread and common phenomenon in fungi functioning to improve the capacity of fungi to survive in association with heterotrophic organisms.

PLANT-FUNGUS INTERACTIONS

Fungi present a universal characteristic which is the impossibility of doing photosynthesis, hence they are completely heterotrophic. Also, fungi must acquire their nutrients from the environment, from living, dying, or dead organisms. They can attack many different hosts so virulently that they kill the hosts and then absorb the released nutrients. In addition, some fungi are able to develop a symbiotic relationship with their hosts varying from mutualism to parasitism.

Symbiosis between a fungus and a plant is a widespread phenomenon in nature, which has been prevalent, ancient, evolutionary consequential in promoting diversity, and plays a major role in structuring plant communities by affecting colonization, competition, coexistence and soil nutrients dynamics. Most, if not all, plants studied in natural ecosystems are infested by fungi without none external manifestation of disease. By definition, those fungi are called endophytes and occur in an astonishing diversity.

The term 'endophyte' has been controversial since its appearance and there are various reports that endophytes can become parasites under certain conditions. Since there are lots of independent evolutionary events, the distinctions between mutualistic and parasitic fungi are unavoidably blurred. However, a good number of published studies are still based on the conventional wisdom that endophytic fungi are plant-defending mutualists. It is known as mutualism, the interaction that is beneficial for both partners, being of foremost ecological and evolutionary importance.

Into this context, it is necessary to have a range of fungus-plant adjustments to establish a balanced mechanism among, for example, the endophyte and host reproduction conflicts, energetic costs of harboring the endophyte, the nutrient demand of the endophyte, the suppression of the host immune systems that not only allow the endophyte to subsist but also might increase the susceptibility to other fungal pathogens, and host control of fungal growth within the plant.

Therefore, this mutualistic relationship has been evolved during the great physical proximity between the interacting cells as a finely tuned balance of antagonisms with a great expenditure of energy to maintain a stable state that disadvantages neither partner. Under this view, disease is the exception in plant-microbe interaction, being regarded as an unbalanced status of a symbiosis. Currently, key elements for the evolution of the endophytic lifestyle involve multiple levels of causation, the whole network of species that interact with host plant directly or indirectly (multispecies interaction), are swayed by biotic and abiotic environmental conditions, influenced by genetic factors, and have traits related to mutual exploitation. The endophyte profits from host plant by receiving organic nutrients, shelter and guaranteed transmission to the next host generation whereas infected plants are more vigorous, drought resistant and toxic to herbivores, nematodes and pathogens than uninfected plants, as has been shown for the symbiotic interaction between *Neotyphodium* endophytes and certain grasses resulting in the synthesis of a number of alkaloids which defend the plants against herbivory.

Sometimes this gain is extended to a third entity of the symbiosis, for instance a hemiparasitic plant. In this case, the endophyte also enhances the resistance of the hemiparasitic plant, which has an ameliorated performance that means a reducing in the growth of the host grass; straight away the mutualistic endophytic fungus becomes parasitic in the presence of the hemiparasitic plant.

Further, the mutualistic association can suffer modifications in the chemical balance. It happens in the case of *Brachiaria plantaginea,* a weed grass species. It yields aconitic acid that stimulates the development of the endophytic fungus *Fusarium solani* into the seeds of other weed species competitor. However, the stimulation of that fungus has an adverse effect on germination of those competitors.

The molecular and biochemical basis for the switch from endophytic to parasitic lifestyle are characterized by an imbalance in nutrient exchange that can explain why colonization of different hosts can cause a fungus to adopt contrasting lifestyles. In this regard, *Colletotrichum magna,* a fungal pathogen which provokes anthracnose in cucurbit plants, exercises an endophytic lifestyle when growing asymptomatically on an assortment of non-cucurbit species.

As a result of these long-held interactions, it is feasible to arouse that some endophytic microbes may have devised genetic systems allowing for transfer of information between themselves and the host plant. As a consequence, the associated microorganisms can learn as much as teach biochemical pathways in order to produce substances common to their hosts or *vice versa* that may have application outside the host plant in which they normally reside. In fact, the study of natural products from plants and their endophytes has shown the probability of that hypothesis.

From the toxic locoweeds *Astragalus mollissimus, Oxytropis lambertii* and *O. sericea* were isolated endophytic fungi capable of producing the alkaloid swainsonine, also produced by their hosts. This compound causes locoism, a disease of ruminant animals. 3-Nitropropionic acid, a potent antimycobacterial agent, was isolated from endophytes and several plants species. Interestingly, some useful plant-derived anticancer drugs have also been identified from endophytic fungal cultures. Paclitaxel (Taxol©), one of the most important drugs available for the treatment of breast and ovarian cancers, was isolated from *Taxomyces andreanae,* an endophytic fungus associated with the Pacific yew *Taxus brevifolia.*

An endophytic fungus from leaves of *Catharanthus roseus* was reported to biosynthesize the potent antileukemia agent vincristine. *Entrophospora infrequens,* an endophyte isolated from *Nothapodytes foetida* (Icacinaceae), was able to produce camptothecin, chemotherapeutic agent efficient against lung, ovarian and uterine cancer, which was first isolated from *Camptotheca acuminate* (Nyssaceae). Podophyllotoxin, a natural product precursor of useful anticancer agents, was found to be synthesized by *Trametes hirsute,* a novel endophyte from *Podophyllum hexandrum,* and also by the endophytic fungus *Phialocephala fortinii,* associated with *Podophyllum peltatum.*

The structures of the compounds 1-6. Based on these examples, knowledge of plant-microorganisms interactions can direct the research of novel bioactive natural products for pharmaceutical and agrochemical industries.

In addition to the previous examples, a significant number of interesting molecules has been found to be produced by endophytes, including alkaloids, steroids, terpenoids, isocoumarins, quinones, flavonoids, phenylpropanoids, lignans, peptides, phenolics, aliphatics, and volatile organic compounds, many of them biologically active. Therefore, endophytes are an outstanding source of small molecules.

THE PLANT AS A HOST FOR FUNGI

The most readily and widely available source of organic carbon is the plant. However, while all plants photosynthesise or have access to photosynthates, plants are not a homogenous or uniform group of organisms. As well, plants have evolved a suite of defences against unwanted invaders. Thus, though very attractive to saprotrophs, plants are a highly diverse kingdom, with different approaches to their interactions with fungi.

In the first instance, plants have quite characteristic defences. The defences have been categorised as generalised, including tannins and lignin, and induced, including phytoalexins. Further, the nature of the defence appears to be related to the mineral nutrients available to the plant. Carbon-based defences are more common in poor soils while nitrogen-based defences are more frequently found in plants growing in rich soils. Thus we might ask why plants become colonised by fungi at all.

The fungus removes organic carbon from the plant. It might be argued that this loss is tolerated by the plant because the plant grows in sunlight and therefore can afford to pass some of its photosynthate to the second party. Others would argue that loss of organic carbon would only make sense if the plant benefited in some way from the association. The plant has three quite clear types of association with fungi: endophytes of the leaves, stems and maybe roots, mycorrhizal fungi of the roots, and plant pathogens *sensu lato*. Note that epiphytes are not considered as a separate group here.

SHOOTS

A clear relationship of benefit to both has been demonstrated by two groups of endophytes,vertically transmitted fungi such as Neotyphodium in grasses, and horizontally transmitted fungi such as Rhabdocline parkeri in Douglas fir. In general, the fungus has a complex interaction with the host. The fungus lives for some years within the leaf tissue and gains access to organic carbon of the host.

The host uses a diverse array of defences which appear to prevent the endophyte spreading through the tissue. Plant responses are apparent and in some cases, the fungus is constrained to one or two cells until the tissue senesces. The fungus benefits the host in both the examples above. The host has reduced insect herbivory in the presence of the endophyte. The fungi produce secondary metabolites which can slow rates of metabolism of those insects that feed from the plant. Some further metabolites deter insects from landing and feeding. Overall, the colonised host plant gains a competitive advantage over uncolonised plants.

The fungus gains access to an almost unlimited supply of organic carbon. The host provides a protected and predictable environment. In the case of Neotyphodium, the host also provides the mechanism for initiation of colonisation of the next plant generation.

Quite clearly, the association is mutually advantageous in these cases. A wide range of fungi can be isolated from within healthy plant tissues. The cost to the host in these cases (endophytes sensu lato) is unknown. The benefit to the host is unknown. Benefit may be in the form of a reduction in herbivory. Alternatively, the endophyte may compete with potential pathogens, and other members of the general biota of plants to the benefit, or detriment, of the host plant.

ROOTS

A clear relationship of mutual benefit has been demonstrated in the case of mycorrhizal associations. In mycorrhizas, fungi colonise young roots and access organic carbon from the host. In all but orchid mycorrhizas the fungus supplies the host plant with increased mineral nutrients, commonly phosphate and/or nitrogen from soil. In the case of orchids, the precise physiological relationship is unclear because the fungus germinates orchid seed in the wild, supplying organic carbon to the plant. The fungus gains no apparent benefit, though density of mycorrhizal fungi declines dramatically away from the surface of the plant.

While many fungi can be isolated from leaves, fewer are present in the living part of the root. Many saprotrophic fungi exist outside the exodermis in young roots, and outside the endodermis once cortical cells have senesced. However, these fungi do not overcome host barriers, are living in dead or dieing tissue, and so are not equivalent to endophytes in leaves.

On some occasions, the number of fungi inside the living zone of the root increases. Water and nutrient stress to the host plant and absence of mycorrhizal fungi appear to be associated with increased colonisation. These fungi are probably general pathogens which have entered because stress has reduced host defences. The presence of competitors of pathogens has also been documented. Binucleate Rhizoctonia appears to competitively reduce colonisation of pathogenic Rhizoctonia solani. Similarly, pathogenic and saprotrophic species of *Fusarium* interact. However, apart from mycorrhizal fungi few root endophytes have been found to benefit the host plant.

PATHOGENS

Interestingly, most potential pathogens do not penetrate the barriers of the host plant. We see pathogens in plants, but only a few of the enormous array known to cause disease. In a few cases, the fungus may penetrate the host tissue, only to have it collapse around the penetration tube. Plant penetration is the exception, rather than the rule. Evolution of pathogen/plant interactions appears to have led to a situation where the plant is not seriously disadvantaged in natural ecosystems.

However, in cropping systems the interaction can be devastating. When the pathogenic fungus initiates infection, it spreads rapidly. This is because the crop is a genetically uniform target (monoculture), growing in uniform environments, at the same rate. The pathogen has the genetic constitution to avoid recognition by the host, and the physiology to parasitise the entire crop. Some of the most spectacular crop losses have been where a single crop has been almost wiped out by a single pathogen. While these incidents have historical importance, they also contain lessons for plant breeders and growers. The ramifications include appropriate targets for plant genetic manipulations, attitudes to sources and use of genetic diversity, and reliance on single genotypes.

COLONISATION

In all these cases, the interaction between fungus and plant has involved a series of steps between plant and fungus. When the spore lands on the plant surface, it releases a chemical message which may be recognised by the plant. The spore attachs to the surface, germinates, and the emergent germ tube then forms an appressorium. A fine penetration peg grows through the plant surface, either directly through the epidermis, or via a stomatal opening. The hypha then ramifies through the plant tissue.

Initial energy for colonisation comes from the spore, perhaps with a small contribution from plant exudates on the surface. This leads to the collapse of the spore and the appressorium. Plant carbohydrates become increasingly important as the tissue is penetrated.

Presumably, those fungi that are recognised as pathogens will induce an immediate response to deter colonisation. The details of pathogenesis are now understood. The details of interaction between mutualistic symbionts are still being clarified, though may well be similar. It is clear that colonisation is an active process which involves a series of steps enabling colonisation and yet controlling spread. Specific fungi interact with each plant species. A small number of generalists also establish colonies in a range of plants. Presumably, the process of recognition and colonisation is under tight genetic control, and may be as complex as the immune system in mammals.

AQUATIC FUNGI

Aquatic fungi, in the broadest sense, include fungi present transiently in water, terrestrial fungi that release spores which are dispersed in water, and species that function entirely within water. Water is found in transient pools after rain, swamps, estuaries, permanent rivers and oceans. Within a single environment, the water can have predictable characteristics. At the edge however, water bodies may contain soil and plant materials, and will be a changing enironment.

Ingoldian Fungi

The major input of plant materials in streams comes from leaf fall. These materials contain a range of endophytes of terrestrial fungi. However, the litter soon becomes colonised by a characteristic group of hyphomycetes with specialised spores. The fungi are known as Ingoldian fungi. They colonise the litter and replace the terrestrial fungi. Further, a typical succession takes place, analogous to succession found in compost and terrestrial leaf litter.Like the common terrestrial fungi, the Ingoldian fungi include many species that appear to lack specificity to the host plant; they are also found around the world.

Other Types

Ingoldian fungi are not the only fungal inhabitants of litter. Another group form large coiled or rounded conidia which trap air, and these are known as aero-aquatic hyphomycetes. The advantage of their spores is that when the water dries, spores can lodge in the mud. On wetting again, the conidia float away on the water, being available to colonise leaf material when they make contact. Colonisation is rapid, and sporulation may take place within 7 days. The hydrophobic conidia are formed on aerial conidiophores, which on release, float on the surface of the water.

The Chytridiomycota can only disperse in water. The zoospore requires at least a film of water to move away from the sporangium. They are readily trapped from bodies of water, and are also found in soil, or on surfaces where a water film forms transiently, on a regular basis.

Distribution of Spores

Ingoldian fungi have a number of interesting features. While the spores are characterised by a set of similar morphological features, it turns out that the teleomorphs are in a diverse range of both Ascomycota and Basidiomycota. The fungi can enzymically degrade the litter and are probably far more common on the leaves than bacteria, at least in the early stages of decomposition. The fungi form conidia while ever leaf litter is available. In streams, seasonal peaks of spores in late summer will be found in the south, and in the late winter in the north, coinciding with the main season of leaf fall. However, the spore season is much longer and less pronounced where the leaf fall is less seasonal.

Many of the fungi also colonise roots of trees on the stream banks. The fungi form sexual spores that lack any features enabling dispersal in water. In fact, the sexual stages may enable dispersal from stream to stream, with the asexual stages increasing colonisation within a stream. After all, release of conidia into a stream inevitably leads to dispersal downstream, not upstream. Further, as the source of leaf litter is highly variable, a source of propagules unrelated to the litter would enable faster build-up of inoculum in the stream.

Interactions with Invertebrates

The size of the population of Ingoldian fungi is also controlled by invertebrates. These animals graze on leaves, preferring colonised leaves, presumably because they have higher protein, lipid and mineral phosphate content due to the fungi, and because the leaf has been markedly predigested. Invertebrate competition markedly increases the rate of litter breakdown, and litter usually disappears rapidly from streams. However, it is argued that the presence of fungi is essential to the invertebrates because the fungi increase the nutritive value, providing essential nutrients not available in leaf material.

FUNGAL CYTOPLASM

The cytoplasm and most organelles and inclusions of fungal cytoplasm are typical of eukaryotic organisms:

- *Nuclei:* Nuclei are always present in living cells. In the fungi nuclei are 1 - 3 μm, somewhat smaller than most other eukaryotic organisms, where they range from 3 - 10 μm. The appearance is similar, though the amount of DNA is generally less, even allowing for the haploid state of most nuclei. Nuclei may move through the cytoplasm of most filamentous fungi, even through the dolipore septum of the Basidiomycota.

- *Mitochondria:* The mitochondria of fungi are clearly recognisable. They have a double bilayer membrane and contain complex internal membranes. They differ from other eukaryotic organisms in that the mitochondria are commonly elongate, oriented along the hyphal axis. The membranes are organised as parallel lamellae usually oriented along the long axis. This orientation is particularly common in older regions of the hypha where vacuoles comprise a large proportion of each compartment, and the cytoplasm is between the vacuole and the wall. The morphology of mitochondria in yeast cells may differ. Giant, branched mitochondria have been observed in yeasts, and intermediate forms occur in cells transforming from yeast to filamentous growth.
- *Cytoplasm:* The cytoplasm is typical is all respects of a eukaryotic cell. Of particular interest is the presence of plasmids. These have been characterised in yeasts. As many as one hundred plasmids are found in yeast cells. Plasmids are also found in filamentous fungi, where some are associated with disease virulence.
- *Membranes:* The organelles are associated with an internal membrane system, in some cases linked in with parts of the membrane. This seems particularly important for movement of materials about the cell, from the site of formation to the site of delivery. The components include vesicles and tubules.
- *Tubules, Vesicles and Vacuoles:* Fungi are characterised by the presence of vacuoles of a spherical to tubular shape. Tubular vacuoles pass along the hyphae, through pores of the septa, interconnecting the thallus. The discovery of tubules is relatively recent. Previously, use of chemical fixative caused tubules to fragment into vesicles. Use of fluorescent markers and microscopic examination of living hyphae have enabled the tubules be observed. Tubules are most concentrated in the hyphal tip. It appears that tubules swell into localised vesicles. The vesicles are moved rapidly along the tubule, through pores of the septum, and to or from hyphal tips. Each compartment may have many tubules operating, at different speeds, and in different directions.

  In older hyphae, the principle component of compartments appears to be the vacuole. The vacuoles appear to have several functions. Recent evidence suggests the vacuoles may have considerable activity and dynamism. They may be involved with recycling, storing and transporting. Vacuoles appear to store enzymes, other macromolecules, lipids, mineral nutrients such as polyphosphate, and toxins. The control of pH and ion homeostasis in compartments appears to rest within the vacuole. Thus vacuoles have a critical function in fungi.

Because vacuoles and the tubular membrane system are linked, it has been postulated that fungi manufacture growth requirements in vacuoles, and deliver growth requirements to other parts using the tubular system. In growth, wall components may be manufactured in vacuoles and delivered to the tip, where the wall is laid down. The wall components would include wall precursors, and enzymes necessary for wall growth. At the same time, enzymes for extracellular digestion would be delivered from the endoplasmic reticulum to the tip, and the results of digestion transported from the tip to the rest of the thallus, presumably including sites of storage.

- *Other Organelles:* A number of organelles appear in specific groups. The largest division of fungi are the Ascomycota. Filamentous forms always have an inclusion called the Woronin body that is associated with the simple pore of the septum. Woronin bodies are spherical electron dense bodies approximately 0.1 to 1 mm usually seen with surrounded by an electron light halo in EM preparations. The contents are probably protein, surrounded by a membrane. While their function is unclear, Woronin bodies appear to block pores following damage to compartments.

Finally, some vesicles called lomasomes are located between the plasmamembrane and the wall. The function of lomasomes is unclear. Because they are always located in zones of active hyphal elongation, a role in polymer synthesis and deposition has been suggested.

Anaerobic Fungi

- *Hydro Genosomes:* In some obligately anaerobic organisms, including fungi, an organelle is observed which appears to produce hydrogenase and pyruvate oxidoreductase. The enzymes function in the anaerobic conversion of organic carbon to energy. The organelle is called a hydrogenosome. Hydrogenosomes have been observed in the Chytridiomycota found in the rumen of herbivores, where mitochondria are apparently absent.

Polarity

Fungal compartments have a decided differentiation from one end to the other. The differences at the tip are most pronounced. Walls are thinner and less well differentiated. Structures associated with delivery of materials for elongation are most obvious and active. In addition, the concentration of Ca^{++} is higher at the tip. Associated with physical structures and chemical signals are electrical fields. While they differ along the hyphal tip, an applied field may alter tip growth rates and directions. The overall evidence suggests these are a consequence of tip regulation rather than the cause. Finally, turgor pressure appears to be associated with hyphal elongation, suggesting that the movement of ions and water into the cell is highly regulated, and part of the explanation of polarity.

NUCLEAR STATES

Each fungus can be found in a variety of nuclear states. That is, the number of nuclei (ploidy) and the degree of duplication (haploid/diploid) can vary. This is because the life cycle of the fungus may have significant periods in each part of the life cycle. The effect of the variation in ploidy is quite significant. If fungi exist as haploids for significant periods, then selection against specific alleles will ensure those which are deleterious will be removed. Thus subsequent diploids will have only neutral or beneficial alleles for that environment.

Cell Cycle

A haploid organism contains a single set of chromosomes in each of its nuclei. Following plasmogamy (fusion of cytoplasm) in filamentous fungi, the number of nuclei doubles, though these need not necessarily go through karyogamy (fusion of the nuclei) immediately. The state where two genetically distinct haploid nuclei coexist in one cell is called a dikaryon. When the nuclei fuse in karyogamy, the nucleus becomes diploid. In this cycle, fungi are similar to most organisms. However, the fungi almost always then pass immediately into meiosis, resulting in the formation of spores with a single haploid nucleus.

Variations

However, the fungi have a number of variations on this theme which provide some interesting advantages. Some fungi lack septa in their hyphae. Thus, the fungus is essentially multinucleate, and the nuclei are usually haploid. The nuclei may be of the same or compatible and different types. The thallus of Zygomycota is non-septate, and multinucleate. The diploid state is usually only found in the sexual zygospore.

Some fungi have significant periods when they are haploid. The vegetative states of most Ascomycota are haploid. The fungus goes through plasmogamy and passes immediately to karyogamy, followed by meiosis and the formation of ascospores. These fungi are only diploid for very short periods of time, and only in very limited parts of the thallus.

Among the Basidiomycota, a variety of nuclear states in compartments has been reported. Spores germinate to form a haploid mycelium. Plasmogamy with a compatible haploid mycelium results in a dikaryotic mycelium, where both nuclei may be functional. The dikaryotic mycelium may never pass into the diploid state. However, if it does, then the sexual phase may be reached, where karyogamy might be followed by meiosis, though these may be separated in time. Reports of multinucleate mycelia indicate that more than two haploid nuclei may coexist in a cell. Indeed, haploid and diploid nuclei may coexist in one compartment.

The nuclear state of most fungi is regulated. One species is commonly found with one nuclear state. This is particularly evident in fungi with secondary homothallism. The fungus can mate with itself in the absence of a compatible mating type. The variation from common behaviour is possible because nuclei may move from compartment to compartment, and the expression of mating type need not necessarily be found at one locus. Indeed, nuclei can function independently of one another. Following formation of a compatible dikaryon, the incoming nucleus passes through mitosis independently of the second nucleus, with one of the resultant nuclei passing through to the next compartment.

FUNGAL DISEASES

Fungal diseases are called mycoses and those affecting humans can be divided into four groups based on the level of penetration into the body tissues:

1. Superficial mycoses are caused by fungi that grow only on the surface of the skin or hair.
2. Cutaneous mycoses or dermatomycoses include such infections as athlete's foot and ringworm, in which growth occurs only in the superficial layers of skin, nails, or hair.
3. Subcutaneous mycoses penetrate below the skin to involve the subcutaneous, connective, and bone tissue.
4. Systemic or deep mycoses are able to infect internal organs and become widely disseminated throughout the body. This type is often fatal.

The most common type of subcutaneous mycosis seen worldwide is sporotrichosis, which occurs most often in gardeners and farmers who come in direct contact with soil. This is a chronic infection caused by the fungus *Sporothrix schenckii,* occurring in three forms. The 'cutaneous lymphatic form' is characterized by a single pustule or nodule that forms at the site of invasion.

This is followed by lymphatic spread and the development of numerous subcutaneous lesions. This 'disseminated form' is marked by multiple, painless cutaneous or subcutaneous nodules, which can form into ulcers or abscesses involving the muscles, joints, bones, eyes, gastrointestinal system, mucous membranes, and nervous system. The 'pulmonary form' results from the inhalation of spores, but produces much the same forms of the disease.

Other forms of subcutaneous mycoses occur mostly in the tropics and subtropics and are caused by several fungal species. These conditions are called chromomycosis (producing wartlike nodules that can ulcerate) and maduromycosis (or mycetoma — a chronic slowly progressive destructive infection involving several layers of skin, producing abscessing granulomas). Treatment is difficult and often requires surgical removal of the offending tissues.

Dermatomycoses is a superficial fungal infection that penetrates only the epidermis, hair, or nails. About thirty different species of the genera *Epidermophyton, Microsporum,* and *Trichophyton* (collectively known as dermatophytes) cause infections commonly known as athlete's foot, jock itch, and ringworm. The term tinea, along with the area of body involved, is also used when referring to these infections. Tinea is Latin for worm or grub because the infections were originally thought to be caused by worm-like parasites.

For example:

1. Tinea capitis is an infection of the scalp common in school children;
2. Tinea corporis is an infection on the body such as ringworm;
3. Tinea pedis occurs on the feet especially the classic athlete's foot;
4. Tinea cruris is an infection of the genital area commonly called 'jock itch;'
5. Tinea barbae affects the area of the beard;
6. Tinea unguium or onychomycosis is an infection of the toenail or fingernail;
7. Tinea manvum affects the hands.

Different species afflict different anatomical sites in different ways. Dermatophyte fungi typically causes scaly lesions that tend to be circular (ringworm). Dermatophyte fungi are molds, but there are yeasts that also cause skin infections. *Trichophyton rubrum* and *T. mentagrophytes* are the dermatophytes that cause athelete's foot.

Malassezia furfur causes a type of dandruff, as well as an infection known as *Pityriasis versicolor* (formerly *Pityrosporum ovale* and*P. orbiculare*). *M. furfur* is actually a complex of seven species that are normally harmless, but, for some reason, turn pathogenic at times.

Infection produces a flaking skin rash and either a bleaching of the skin or a deposit of more pigment to the area. Lesions are generally found on the trunk, and sport a golden yellow florescent colour when an ultraviolet light is shone on them. Systemic mycoses occur in two basic forms — respiratory and disseminated tissue. If left untreated, the disseminated form is usually fatal. Surgical removal of large pulmonary lesions have been useful in some cases.

Other systemic mycoses include:

1. Coccidioidomycosis (valley fever) is caused by *Coccidioides immitis,* a fungus found in desert areas and prevalent in the southwestern US and in some areas of Central and South America. These spores are carried by desert winds and infect the pulmonary spaces. As many as 80% of the population of the central valleys of California test positive for this fungus.

2. Histoplasmosis is caused by the fungus *Histoplasma capsulatum*, which grows in soil fertilized by bat droppings. It is most prevalent in the Ohio River and Mississippi River valleys, where positive skin tests reveal that at least 80% of the population have been infected. Mild lung infections are usually the result of the disease and often go unnoticed, but, on healing, small calcified nodules that are often mistaken for tuberculosis lesions on chest X-rays are left on the lungs.
3. Blastomycosis is caused by the fungus *Blastomycoses dermatitidis*. It grows in soil as a mold producing conidia that infect human by the airborne route. Along with common mild respiratory infections, skin lesions can also appear. The fungus grows as a budding yeast in human tissues.
4. Cryptococcosis is caused by the yeast *Cryptococcus neoformans*, normally found in soil and habitats of pigeons. It exists only in the yeast form. Scientists speculate that it is millions of years old. Most infections are of a mild respiratory nature, but, in compromised persons, the infection may spread throughout the body affecting mostly the central nervous system. *Cryptococcus neoformans gattii* is usually found in tropical regions, but has now made an appearance on Vancouver Island in Canada, with several deaths attributed to it.

The fungus, *Coccidioides immitis* or cocci, for short, is an infectious, but not contagious, disease contracted by inhaling spores produced in soil. Anything that disturbs the soil will send these spores into the air — construction, earthquakes, off-roading, dust storms, archeological digs, etc. So far, it is found only in the Western Hemisphere, particularly in the American southwest.

The strong winds of southern California between Thanksgiving and Christmas produce numerous cases of these deadly cocci illnesses. The fungii grow so rapidly that death can result very quickly. Once inhaled, the spores often migrate to one area and lay dormant in a pocket of tissue, looking like a hanging beehive. The only known drug proven effective has been amphotericin B, but it has horrendous side effects, killing healthy human cells as well as the invasive agent. This leaves the kidneys and bone marrow especially sensitive, and usually results in kidney damage and anemia.

Mucormycosis refers to the fairly rare diseases produced by a variety of common fungi of the order Mucorales. These, too, are seen only in the severely immunocompromised patient. The fungi penetrate the respiratory or intestinal mucosa and can enter through breaks in the skin as well. Localized lesions may develop, followed by a spreading to the blood, which carry it to all organs. Death often results. Fungi most commonly associated with specific immunocompromised patients:

1. *Candida species, Aspergillus* species, *Phycomyces* species (Leucopenia or bone marrow failure);

2. *Candida, Cryptococcus, Coccidioides, Histoplasma* (Cellular immunity or tissue transplants);
3. *Zygomyces, Rhizopus, Mucor, Absidia* (Diabetes);
4. *Zygomyces* (steroid therapy);
5. *Candida, Cryptococcus, Histoplasma* (malignancies as in leukemia and lymphoma, and Hodgkin's disease);
6. *Candida, Cryptococcus, Histoplasma.* (AIDS)

Clinically significant fungi and the areas of disease they cause:

1. *Malassezia furfur* and *Exophiala werneckii* (superficial skin)
2. *Piedraia hortae* and *Trichosporon beigelii* (hair)
3. *Microsporum* species, (skin and hair)
4. *Epidermophyton* species (skin and nails)
5. *Trichophyton* species (skin, hair, and nails)
6. *Sporothrix schenckii, Cladosporium* species, *Phialophora* species, and *Fonsecaea* species (subcutaneous/lymphatic tissues-chromoblastomycosis)
7. *Histoplasma capsulatum, Coccidioides immitis, Fusarium* species, *Penicillium* species (systemic respiratory)
8. *Blastomyces dermatitidis* (subcutaneous/respiratory)
9. *Cryptococcus neoformans* (respiratory/CNS)
10. *Aspergillus* species, *Mucor* species, *Candida* species, and *Rhizopus* species (opportunistic involving various body sites).

FUNGAL DISEASES: DAMPING OFF–CROWN AND ROOT ROTS

Plants in seedbeds may be diseased, either before or after their emergence from the soil, and the disease is called pre- or post-emergence damping off, respectively. In the first case, seedlings do not emerge in patches of the seedbeds. In the second case, plants rot quickly and drop down on the soil. Low temperatures and very wet soils, which delay the growth of the plants, favour infection.

A large number of fungi may cause damping off, but *Pythium* spp., *Phytophthora* spp., *Fusarium* spp. and *Rhizoctonia solani* Kühn are the most common. Nowadays, due to the use of improved technology, damping off is no longer a severe disease in greenhouses. However, root rots and crown rots are still destructive in soil, though not in soilless cultures. The most widespread diseases are as follows.

Pythium and Phytophthora Rots

Various *Pythium* spp. and *Phytophthora* spp. may damage the lower part of tomato, pepper, cucumber, carnation, poinsettia, gerbera, etc. both in soil and soilless cultures. In tomato a root and crown rot extending a considerable height above the soil level may occur. The infected area has a dark discoloration and the pith is usually destroyed. *Phytophthora nicotianae* Breda de Haan var. *parasitica* (Dastur) G.M. Waterhouse is the most common pathogen. In pepper a similar disease caused by *Phytophthora capsici* Leonian is very common. Collar, stem and fruit rot as well as leaf spots may occur. In cucumber a soft rot of the young plants at the soil level may occur soon after transplanting.

Infected tissues shrink and in wet weather a white mycelium develops. Infected plants wilt and die quickly. Poinsettia grown in pots also suffers from Pythium rot. Severe root rot, extending above ground in succulent plants, and quick death are the main symptoms. In cucumber and poinsettia, *Pythium ultimum* Trow, *Pythium irregulare* Buisman, *Pythium debaryanum* Auct. Non R. Hesse and *Pythium aphanidermatum* (Edson) Fitzp. are mostly involved. Carnations infected by *Pythium* and *Phytophthora* species develop soft rot at the collar and in the root system, resembling Rhizoctonia stem rot

Corky Root Rot of Tomato

The pathogen damages mostly tomato, but also eggplant, melon, etc. Initially tomato leaves turn dull green and growth is stunted. Later, leaves take on a bronze colour and curl downwards. Necrosis of the leaflets follows. Young roots are brown and poorly developed. Scattered lesions appear on the surface of the larger roots which become corky with cracks of different sizes. Yield may be severely reduced. The pathogen survives on the infected root debris due to the presence of minute sclerotia. It is a cool weather disease. In subtropical countries it progresses during the winter and plants start to recover by early spring.

Rhizoctonia Stem Rot (R. solani)

This infects a large number of plants, such as tomato, carnation, poinsettia, etc. causing symptoms resembling Pythium or Phytophthora rots. Rhizoctonia stem rot is mainly confined to the collar. Carnation is very susceptible. Infected plants show pale brown dry lesions, with circular rings, at soil level. Growth is stunted and leaves become dull green. Complete wilting soon follows. Strands of the pathogen develop on the lesions and stems break easily at the infected area. All *Pythium* and *Phytophthora* species as well as *R. solani* are common soil inhabitants.

They survive in the soil. Infection usually takes place at the time of planting and symptoms appear very soon in Pythium and Phytophthora rot or several weeks later in Rhizoctonia stem rot *Rhizoctonia solani* may infect at moderate soil moisture levels, but *Pythium* spp. and *Phytophthora* spp. infect only in water-saturated soils.

Crown and Root Rot of Tomato

In plastic greenhouses, a yellowing of the lower leaves appears in infected plants during late winter, when many fruits have already set. In severe infections the whole plant becomes chlorotic and wilts. A dry lesion up to 10 cm long appears on part of or all around the collar. There is a brown discoloration on the root system, predominantly at the end of the main root, the base of the stem and the vascular region of the central root A large number of microconidia, which disseminate the pathogen, appear on the infected stem. The fungus survives by chlamydospores which develop in the soil. The disease is favoured by cool weather.

Black Root Rot of Cucurbits

The disease has been recorded in several countries of northwestern Europe and elsewhere. It infects cucumber and melon, causing a brown rot in the cortical tissue of the root system. Soon a large number of sclerotia develop and the infected tissues turn black. Severely infected plants wither and die. Infection is favoured by cool weather. The pathogen survives in the soil for several years by means of sclerotia.

Control

Effective control of the above-mentioned diseases may be obtained selectively by the following means:

- Use of naturally or artificially suppressive substrates;
- Early drenching by effective fungicides;
- Soil disinfestation;
- Use of resistant cultivars (cvs);
- Grafting on resistant rootstocks;
- Biological control

WILTS: GREENHOUSE CROPS

All major greenhouse crops suffer from one or more wilts. In several crops wilts are the main diseases due to the damage they cause and the difficulty of controlling them.

Fusarium Wilt

The most common Fusarium wilts in greenhouses appear on: tomato [*Fusarium oxysporum* Schlechtend.:Fr. f. sp. *lycopersici* (Sacc.) W.C. Snyder & H.N. Hans.], cucumber (*Fusarium oxysporum* Schlechtcnd.:Fr. f. sp. *cucumerinum* J.H. Owen), melon (*Fusarium oxysporum* Schlechtend.:Fr. f. sp. *melonis* W.C. Snyder & H.N. Hans.), carnation [*Fusarium oxysporum* Schlechtend.:Fr. f. sp. *dianthi* (Prill. and Delacr.) W.C. Snyder & H.N. Hans.], gladiolus [*Fusarium oxysporum* Schlechtend.:Fr. f. sp. *gladioli* (L. Massey) W.C. Snyder & H.N. Hans.], cyclamen (*Fusarium oxysporum* Schlechtend.:Fr. f. sp. *cyclaminis* Gerlach) and chrysanthemum (*Fusarium oxysporum* Schlechtend.:Fr. f. sp. *chrysanthemi* G.M. Armstrong, J.K. Armstrong & R.H. Littrell). Wilt, yellowing, chlorosis, drooping (mostly of the lower leaves), stunting and brown discoloration of the vascular bands up to the top of the stem are the dominant symptoms. Wilting of the lateral shoots and large lesions on the lower part of the stem are also common in Fusarium wilt of carnation and melon. All the above *F. oxysporum* formae have more than one race. Each of them infects cvs of one host but may colonize the root system of other plants as well. They survive in the soil for several years, due to the production of thick-walled chlamydospores, but inoculum is reduced over the years. Fusarium wilt in tomato, watermelon, carnation, cyclamen, chrysanthemum and gladiolus is favoured by higher temperatures than Fusarium wilt in melon.

Verticillium – Phialophora Wilt

This infects a huge number of plants and among them the majority of the plants grown in greenhouses. It is more severe in Solanaceae such as tomato, eggplant and pepper. Of the floral crops, chrysanthemum seems to be more susceptible. Symptoms are very similar to those of Fusarium wilt. Verticillium wilt is favoured by moderate temperatures. *Verticillium dahliae,* which is more common, survives in the soil for many years due to the abundant production of black resistant microsclerotia, while *V. albo-atrum* survives by producing dark dormant mycelium. A similar wilt caused by *P. cinerescens* damages carnations in several areas.

DESTRUCTIVE OF SEVERAL GREENHOUSE CROPS

Powdery mildews are very destructive of several greenhouse crops.

The following are some of the powdery mildew fungi which most attack greenhouse-grown plants:

- On cucurbits *Sphaerotheca fusca* (Fr.) Blumer. [= *Sphaerotheca fuliginea* (Schlechtend.:Fr.) Pollacci], *Erysiphe cichoracearum* DC. and *Leveillula taurica* (Lév.) G. Arnaud (only on cucumber);
- On solanaceous plants L. *taurica* and *Oidium lycopersicum* Cook and Massee (only on tomato);
- On roses *Sphaerotheca pannosa* (Wallr.:Fr.) Lév.;
- On begonia *Microsphaera begoniae* Sivan.;
- On gerbera *E. cichoracearum.*

Powdery mildew fungi, except for *L. taurica,* may attack all green tissues. Initially, white powdery spots, which enlarge and coalesce to cover large areas, are the dominant symptoms. *Leveillula taurica* infects only leaves.

Light yellow or yellow-green spots on the upper leaf surface, which later become brown, and scarce white mould on the lower surface are the main characteristics. Powdery mildew-infected plant parts may be chlorotic and distorted.Premature defoliation and poor growth are common features of severely infected plants. Infections take place by conidia. Under favourable conditions powdery mildew progresses rapidly. By the end of the season some powdery mildew fungi, such as *S. fuliginea, E. cichoracearum,* etc. may develop cleistothecia with ascospores, but these do not play an important role in the epidemiology of the disease. Conidia are mostly discharged and transferred by wind currents. Animal pests may also disseminate conidia in greenhouse crops. Young conidia readily germinate on plant surfaces depleted of nutrients.

The relative humidity (RH) favouring infection by powdery mildew fungi and development of the disease differs from species to species. For instance, high RH is more favourable for *S. fuliginea* than for E. *cichoracearum.* Therefore, the first fungus is more frequent on greenhouse cucurbits than the second. High RH may favour spore germination of powdery mildew fungi, but free water may be deleterious. RH at 97–99% is optimal for spore germination of *S. pannosa* and *S. fuliginea.*

At RH below 75% spores of *S. pannosa* do not germinate, but mycelium development and sporulation may occur at RH as low as 21–22%. Powdery mildew fungi overwinter on cultivated plants or weeds, which survive in or outside the greenhouse. Chemicals such as demethylation inhibitors (DMIs) (triadimefon, fenarimol, etc.), pyrimidines (ethirimol, bupirimate, etc.), pyrazophos and dinocap remain the main means of controlling powdery mildews in greenhouses. Biological control agents have also been effectively tested against *S. fuliginea* and *S. pannosa.* Finally, fully resistant cvs of melon and partially resistant cvs of long-type cucumber are available.

Botrytis Diseases

Lind and Botrytis Gladiolorum

Timmermans are *Botrytis* spp. that most damage greenhouse crops. *Botrytis cinerea* causes grey mould on a large range of hosts, including nearly all the major greenhouse plants. All plant parts at different growth stages may be damaged. Due to the diversity of the infected plant parts, several types of symptoms appear on one or on various hosts. On young stems, leaves, flowers and fruits, initially water-soaked spots occur, which rapidly enlarge under favourable weather conditions.

In tomato fruits green-white circular spots called "ghost spots" also appear. On hard plant parts, such as stems and collars, *B. cinerea* causes cankers and parts above them may die. These symptoms are very common on vegetables such as tomato, eggplant, pepper and cucumber. Infected tissues die soon and a grey mould which consists of conidiophores with clusters of spores develops on their surface. In plants, like tomato, black sclerotia develop inside the infected stems. *Botrytis cinerea* also causes very characteristic collar rot in lettuce. The infected plants usually develop large brown necrotic lesions on the stem near the soil surface and the lower leaves. The infection gradually progresses upwards. Infected plants may wither and die in a short time.

Botrytis Tulipae Causes Tulip Fire Blight

Spots of various types on leaves and flowers, lesions on the stem, blossom blight and bulb rot are the dominant characteristics. *Botrytis gladiolorum* damages gladiolus and some other Iridaceae. Large spots on leaves and the stem, pinpoint spots on the flowers, neck rot and soft rot of corms are the most common symptoms. *Botrytis* spp. also infect all types of propagating material, which are either destroyed before planting out or become weak plants which may die before or after emergence. Finally, *Botrytis* spp. may cause severe post-harvest losses in plant products during storage or transportation.

Botrytis cinerea develops and sporulates profusely on any organic material. Spores are disseminated by wind over long distances or by water splashes. Healthy plants are infected through wounds, senescent tissues, directly through the epidermis and rarely through stomata. Symptoms may appear very quickly or infection may remain quiescent and symptoms appear later when tissues age or during storage.

In greenhouses, initial infection depends on spores transferred from outdoors. Later, the inoculum established in the greenhouse is the main source of infection. In plants grown in non-heated greenhouses, low temperature, high RH and low light intensity, prevalent from late November till late March, create good conditions for infection by *B. cinerea*. Botrytis-incited diseases are prevented by ventilation and heating of greenhouses. Fungicides, mostly benzimidazoles and dicarboxymides, are also used extensively.

Nowadays, due to the predominance of resistant strains of the pathogen, they are only marginally effective and growers are advised to combine dicarboxymides with other means of control such as biocontrol preparations.

New fungicides have recently been released, but in greenhouses they are used on a limited scale. Formulations of biological control agents such as Trichodex (*Trichoderma harzianum* Rifai T39) are also available.

Downy Mildews

Downy mildews of tomato [*Phytophthora infestans* (Mont.) de Bary], cucurbit [*Pseudoperonospora cubensis* (Berk. & M.A. Curtis) Rostovzev], lettuce (*Bremia lactucae* Regel), rose (*Peronospora sparsa* Berk.) and snapdragon (*Peronospora antirrhini* J. Schröt.) are the most destructive in greenhouse-grown plants. In tomato, leaves and young shoots are infected first. Fruit infection starts mostly near the stalk and spreads very quickly to the whole fruit. Infected tissues of fruits and shoots are firm and brown. In cucurbits downy mildew appears as yellow, angular or circular spots on the upper surface of the mature leaves of the plant.

Soon the tissues at the centre of the spots die and become light brown. Cucumber and melon are more susceptible than watermelon. Downy mildew of lettuce causes scattered light-green to yellow spots on the upper leaf surface. Old spots become brown and dry up. Downy mildew of rose damages all green plant parts, but leaves are more susceptible. Leaf infection resembles the effect of toxins. Infected leaves have purplish red to dark-brown irregular spots and shed readily. Snapdragon plants infected by *P. antirrhini* are stunted and the top intemodes of the young plants are short. The borders of the lower leaves curl down and then dry. Eventually the entire plant dies.

A white fungal growth (brown for cucurbit downy mildew) on the infected tissues under moist conditions is typical of all downy mildews. Plant infection takes place through stomata and mycelium develops intercellularly. Soon branched conidiophores are produced and protrude through the stomata. Infection progresses in the periphery of the spot which gradually enlarges. Conidiospores of downy mildews are ovoid and hyaline, except for *P. cubensis* which are brown.

They are discharged by hygroscopic changes and disseminate in greenhouses by wind currents and water splashes. Initial infection may take place by spores transferred long distances on the wind. Abundant oospores of *P. antirrhini* develop on dead plant stems. Oospores of *P. sparsa* also very often develop on infected roses, whereas *P. cubensis* and *P. infestans* oospores are rare.

Phytophthora infestans survives on seed potato tubers and spreads to young potato plants after they have been planted. Inoculum is disseminated from potatoes to neighbouring tomato crops. Cucurbit downy mildew can infect all year round several species of cucurbits, grown either in greenhouses or open fields.

There is evidence that *P. sparsa* survives as a dormant mycelium on the infected stems of roses. *Peronospora antirrhini* perennates as dormant thick-walled oospores in dead plant parts and soil. Free water on plant tissues is necessary for downy mildew fungi to cause infection. High RH is also required for good sporulation. *Peronospora antirrhini* is favoured by low temperature and high RH. Free water or high relative humidity is not often a factor limiting downy mildew development in plastic greenhouses. It seems that temperature is more critical.

For instance, *P. cubensis,* with a high maximum temperature for development and infection, may, under certain conditions, infect all year round, whereas *P. infestans* and *P. sparsa* do not infect during the hot period of the year. Downy mildews complete a cycle within about 6–8 days. Thus, under favourable weather conditions they may have several cycles and spread rapidly.

Chemical fungicides remain the major means of control of downy mildews. Dithiocarbamates, chlorothalonil and the systemic phenylamides (metalaxyl, etc.) are the most commonly used in greenhouses. There are some tomato cvs fully resistant to downy mildew and some partially resistant cucumber cvs suitable for greenhouses, but all rose cvs grown for cut flowers are susceptible to downy mildew. Ventilation of the greenhouses may also effectively prevent infection.

Alternaria Diseases

The following diseases, caused by *Altemaria* spp., seriously affect vegetable and floral crops in greenhouses.

Tomato Early Blight (Alternaria Solani Sorauer)

A collar rot of the young plants before or after transplanting may be the first symptom. In mature plants small irregular brown spots, with or without a yellow halo and concentric rings, appear mainly on leaves. Severely infected leaves are ragged and senescent Similar spots without a yellow ring appear along the stem, leaf stalks, pentucles and the calyx.

On fruits, brown to black spots with a leathery surface appear at the stem end. Severely infected plants may be defoliated. *Alternaria Branch Rot and Leaf Spot of Carnation* (Alternaria dianthi *Stev. and Hall.*) This mostly infects carnation cuttings during mist propagation and in wet parts of greenhouses. Small purple spots on the leaves are the first symptoms. Soon they enlarge, and their centre turns brown and then black due to the masses of spores which develop. Stem infection usually appears on the knots.

Alternaria Diseases of Minor Importance for Greenhouses

As well as the Alternaria diseases described above, strains of Alternaria alternata (Fr.:Fr.) Keissl. have been recorded:

- Causing cankers in tomato crops;
- Causing leaf spotting in cucumber;
- Causing mostly post-harvest rotting on tomato fruits.

Also *Alternaria cucumerina* J.A. Elliot may on occasion infect cucumber, melon, watermelon and squash. At present, none of them has any economic impact on greenhouse crops. All *Alternaria* species are facultative parasites mostly infecting weak plants. They survive in the soil on plant debris, but their black spores may also survive on several surfaces in greenhouses. *Alternaria solani* may survive on potato, which is an alternative host. Spores growing on dead material or on host plants are easily disseminated by wind or by splashed water. Plant infection takes place through stomata or directly through leaf surface. Spore germination and subsequent infection take place under a wide range of temperature. RH needs to be higher than 97% for rapid germination, but germination may take place in some cases at RH >75%. Senescent tissues are preferentially infected. The optimal temperature reported for *A. solani* is 18–25 °C and for *A. cucumerina* 20–32 °C. However, temperatures prevailing during the growing period of the respective hosts are not a factor limiting infection.

Control

Alternaria diseases can be prevented by dithiocarbamates, chlorotholonil, iprodione, etc. Hygienic measures and use of healthy propagating material are very important, especially when crops are grown in the soil. Inoculum surviving on plant debris in the soil and spores remaining on the greenhouse frames should be eradicated.

Sclerotinia Rot

This is a common greenhouse disease that damages lettuce, eggplant, tomato, cucumber, pepper, etc. Infection on lettuce begins close to the soil, where a water-soaked area appears. Infection may spread downwards to the roots or upwards to the heart of the plants. Infected leaves fall onto the soil and dry up.

The other plants are infected along the stem, leaves, flowers and fruits. Infected areas become water-soaked. Stem infection is more severe. Leaves above the infection area become yellow, wither and die. In wet weather a white mass of mycelia appears on the infected areas, which gradually develops into black sclerotia. *Sclerotinia sclerotiorum,* which is the most common pathogen, produces sclerotia up to the size of bean seeds, whereas *S. minor* produces smaller sclerotia. Sclerotia fall onto the soil where they can survive for several years.

When weather conditions are favourable they germinate to produce apothecia which release ascospores and cause new infection. High RH and moderate temperature is required for infection. The elimination of sclerotia and the control measures recommended against grey mould are effective against Sclerotinia rot as well.

Didymella Diseases

Two very severe diseases of greenhouse crops are caused by *Didymella* spp.: Didymella stem rot or canker in tomato and eggplant and gummy stem blight in cucurbits. Both diseases damage all aerial plant parts of their hosts in greenhouses when weather is cool and RH high. They may infect the collar and root system causing yellowing and withering of the plants, which may later die. Cankers along the stem and the petioles are also very common. Plant parts above cankers may die.

Both diseases cause large spots on the leaves which may cover the entire leaf surface. Tomato fruits are infected at the stem end. Initially, the infected area is light brown but it soon turns pink due to the large amount of pycnidiospores released. Infected parts may cover one third of the fruit surface. Infection of cucumber and melon fruits by *D. bryoniae* appears mostly at the blossom end. Infection may occur only inside the fruit without being visible on the surface. Soon after infection, a lot of pycnidia appear on the infected areas and their colour turns dark brown.

Dark perithecia also appear a little later than pycnidia produced by *D. bryoniae,* while those of *D. lycopersici* are rare. The inoculum remains in plant residues inside and outside greenhouses. In the first case infection starts through the collar. There is good evidence that infection of the aerial parts by *D. bryoniae* is initiated by ascospores released from infected plant material left outside greenhouses.

In greenhouses the two diseases are rapidly spread by water splashes and cultural practices. Soil disinfestation, destruction of plant residue and strict hygienic conditions delay the outbreak of the diseases. However, disinfested soil is readily reinfested. The fungicides commonly used in greenhouse against other fungal diseases are also effective. Moreover, the reduction of the RH and of free water on the leaf surfaces is very effective.

Rust Diseases

These are a very important group, with many common characteristics. The following are the main rusts affecting greenhouse crops.

Carnation Rust

The disease is more severe on leaves, but other green plant parts are infected as well. Initially, small light green spots appear. They gradually turn to powdery brown blisters due to the herediospores developed. Severely infected plant parts are twisted. Healthy crops are infected by urediospores transferred from neighbouring crops. They are wind or water-splash disseminated and germinate readily on free water. The cycle of the pathogen lasts about two weeks. In greenhouses, where leaves may remain wet for several hours, there may be many disease cycles per crop season.

Rose Rust

The disease is easily identified by the yellow orange rust pustules which develop profusely on the lower surface of older leaves. In greenhouses it is not very destructive.

Several species of *Phragmidium* have been reported to infect rose, but *P. mucronatum* is the most common. It is an autoecious, macrocyclic fungus producing telia by the end of the crop season in the same place as uredospores. They serve as overwintering structures and initiate infection during spring. Free water and temperature 9–27 °C are necessary for the uredospores to germinate.

Chrysanthemum Rust and White Rust of Chrysanthemum

Pale yellow flecks on the leaves followed by dark brown pustules with urediospores are the dominant symptom. Leaves with several pustules may wither and die. No stem infection has been reported. It is a low to moderate temperature disease requiring free water for infection. It survives on infected leaves and is disseminated by wind.

Chrysanthemum white rust is a new and destructive disease of chrysanthemum in Europe and the Mediterranean. Initially, circular white or yellow cushions develop on the lower leaf surface and then soon turn brown. The disease is favoured by high RH and moderate temperatures. Snapdragon rust, geranium rust (*Puccinia pelargoniizonalis* Doidge), etc. are also destructive diseases, but the respective crops are not grown in large acreage. Regular applications of protective fungicides, such as dithiocarbamates and chlorothalonil, or systemic fungicides, such as oxycarboxin and members of the DMIs, are mostly recommended for rust control. Prevention of water condensation is also very effective.

Tomato Leaf Mould

This causes light green to yellow spots on the upper surface of mature leaves. Soon the sporulating fungus growth appears as an olive-green velvety growth on the underside of the yellow spots. The pathogen survives for several months on the greenhouse frame, on the materials used for cropping and in plant debris.

It is disseminated by wind or splashed by water drops. The optimal temperature for infection is 20 to 25 °C. If weather conditions are favourable, leaf mould has several cycles in a season and can destroy the crop completely. There are several races of the pathogen.

Cucurbit Scab

This mostly attacks cucumber, but also squash, melon, etc. It causes nearly circular or angular leaf spots on the leaves, which look water-soaked. Fruit infection is more serious. Initially, water-soaked lesions about 1 cm long, with gummy exudations, develop. A corky tissue usually develops around the lesions, which finally develop a scabby appearance. The pathogen survives on plant debris and spores are air-disseminated. Temperatures of about 15 to 25 °C and RH over 86% favour the disease.

Control

For both diseases, greenhouse ventilation is the best control measure. The disinfestation of greenhouse soil and frames is also very important. Regular application of dithiocarbamates, iprodione, benzimidazoles, etc. are recommended as well. There are several resistant cvs against some races of the pathogens.

6

The Soil Microorganisms and Rhizosphere

THE ECOSYSTEM OF SOIL MICROBIOLOGY

In the biosphere, microorganisms play a key role in maintaining the chemical balance of available nutrients and metabolic waste products. In this way, they help preserve the natural environment. In addition, microorganisms are important in the elemental cycles occurring in the soil.

An ecosystem is the community of organisms found in a physically defined space. In the soil ecosystem, microorganisms are the largest contributors of organic matter. This organic matter is derived from the metabolism of animal and plant waste. The decay-resistant organic matter not recycled combines with mineral particles to form the dark-coloured material of soil called humus.

THE NITROGEN CYCLE

The nitrogen cycle represents one of the most important nutrient cycles found in terrestrial ecosystems. Nitrogen is used by living organisms to produce a number of complex organic molecules like amino acids, proteins, and nucleic acids. The store of nitrogen found in the atmosphere, where it exists as a gas (mainly N2), plays an important role for life. This store is about one million times larger than the total nitrogen contained in living organisms. Other major stores of nitrogen include organic matter in soil and the oceans.

Despite its abundance in the atmosphere, nitrogen is often the most limiting nutrient for plant growth. This problem occurs because most plants can only take up nitrogen in two solid forms: ammonium ion ($NH4^+$) and the ion nitrate ($NO3^-$). Most plants obtain the nitrogen they need as inorganic nitrate from the soil solution.

Ammonium is used less by plants for uptake because in large concentrations it is extremely toxic. Animals receive the required nitrogen they need for metabolism, growth, and reproduction by the consumption of living or dead organic matter containing molecules composed partially of nitrogen. In most ecosystems nitrogen is primarily stored in living and dead organic matter. This organic nitrogen is converted into inorganic forms when it re-enters the biogeochemical cycle via decomposition. Decomposers, found in the upper soil layer, chemically modify the nitrogen found in organic matter from ammonia (NH_3) to ammonium salts (NH_4^{+}). This process is known as mineralization and it is carried out by a variety of bacteria, actinomycetes, and fungi. Nitrogen in the form of ammonium can be absorbed onto the surfaces of clay particles in the soil.

The ion of ammonium has a positive molecular charge is normally held by soil colloids. This process is sometimes called micelle fixation. Ammonium is released from the colloids by way of cation exchange. When released, most of the ammonium is often chemically altered by a specific type of autotrophic bacteria (bacteria that belong to the genus Nitrosomonas) into nitrite (NO_2^{-}). Further modification by another type of bacteria (belonging to the genus Nitrobacter) converts the nitrite to nitrate (NO_3^{-}).

Both of these processes involve chemical oxidation and are known as nitrification. However, nitrate is very soluble and it is easily lost from the soil system by leaching. Some of this leached nitrate flows through the hydrologic system until it reaches the oceans where it can be returned to the atmosphere by denitrification.

Denitrification is also common in anaerobic soils and is carried out by heterotrophic bacteria. The process of denitrification involves the metabolic reduction of nitrate (NO_3^{-}) into nitrogen (N_2) or nitrous oxide (N_2O) gas. Both of these gases then diffuse into the atmosphere. Almost all of the nitrogen found in any terrestrial ecosystem originally came from the atmosphere. Significant amounts enter the soil in rainfall or through the effects of lightning. The majority, however, is biochemically fixed within the soil by specialized micro-organisms like bacteria, actinomycetes, and cyanobacteria.

Members of the bean family (legumes) and some other kinds of plants form mutualistic symbiotic relationships with nitrogen fixing bacteria. In exchange for some nitrogen, the bacteria receive from the plants carbohydrates and special structures (nodules) in roots where they can exist in a moist environment. Scientists estimate that biological fixation globally adds approximately 140 million metric tons of nitrogen to ecosystems every year. The activities of humans have severely altered the nitrogen cycle.

Some of the major processes involved in this alteration include:

- The application of nitrogen fertilizers to crops has caused increased rates of denitrification and leaching of nitrate into groundwater. The additional nitrogen entering the groundwater system eventually flows into streams, rivers, lakes, and estuaries. In these systems, the added nitrogen can lead to eutrophication.
- Increased deposition of nitrogen from atmospheric sources because of fossil fuel combustion and forest burning. Both of these processes release a variety of solid forms of nitrogen through combustion.
- Livestock ranching. Livestock release a large amounts of ammonia

into the environment from their wastes. This nitrogen enters the soil system and then the hydrologic system through leaching, groundwater flow, and run-off.

- Sewage waste and septic tank leaching.

OTHER BIOGEOCHEMICAL CYCLES

In addition to being a site for the nitrogen cycle, the soil is the environment in which several other biogeochemical cycles take place. Among these are the cycles of phosphorus, sulfur, carbon, and oxygen.

THE PHOSPHORUS CYCLE

Living things use phosphorus compounds in the synthesis of nucleotides, phospholipids, and phosphorylated proteins. Phosphorus enters the soil and water as phosphate ions, such as calcium phosphate, during the breakdown of crops, decaying garbage, leaf litter, and other sources. In the phosphorus cycle, microorganisms use phosphorus in the form of calcium phosphate, magnesium phosphate, and iron phosphate. They release the phosphorus from these complexes and assimilate the phosphorus as the phosphate ion (PO_4). This ion is incorporated into DNA, RNA, and other organic compounds using phosphate, including phospholipids. When the organisms are used as foods by larger organisms, the phosphorus enters and is concentrated in the food chain.

THE SULFUR CYCLE

Sulfur makes up a small percentage of the dry weight of a cell (approximately 1 per cent), but it is an important element in the formation of certain amino acids such as cystine, methionine, and glutathione. It is also used in the formation of many enzymes. Many bacteria have an important place in the sulfur cycle in the soil. Sulfate-reducing bacteria grow in mud and anaerobic water environments, where they reduce sulfur-containing amino acids to hydrogen sulfide (H_2S).

Hydrogen sulfide accumulates in the mud of a swamp and gives the environment an odour of rotten eggs. In the cycle's next step, photosynthetic sulfur bacteria metabolize the hydrogen sulfide anaerobically. They oxidize the H_2S, thereby releasing the sulfur as elemental sulfur (S). Elemental sulfur accumulates in the soil. Species of colourless sulfur bacteria, including members of the genera Thiobacillus, Beggiatoa, and Thiothrix, also metabolize the hydrogen sulfide, converting it to sulfate ions, which are then made available to plants for amino acid formation.

THE CARBON CYCLE

Most of the organic matter present in soil originates in plant material from dead leaves, rotting trees, decaying roots, and other plant tissues. Animal tissues enter the soil after death. In the carbon cycle, soil bacteria and fungi recycle the carbon of proteins, fats, and carbohydrates by using the organic plant and animal matter in their metabolism. Without the recycling of carbon, life would suffer an irreversible decline as nutrients essential for life were bound in complex molecules.

The organic matter of organisms is digested by extracellular microbial enzymes into soluble products. Fungi and bacteria then metabolize the soluble organic products to simpler products such as carbon dioxide and acetic, propionic, and other small acids, as well as other materials available for plant growth.

These elements are made available to the root systems of plants. Undigested plant and animal matter becomes part of the humus.

THE OXYGEN CYCLE

In the oxygen cycle, oxygen is a key element for the chemical reactions of cellular respiration (glycolysis, Krebs cycle, electron transport, chemiosmosis). The atmosphere is the chief reservoir of oxygen available for these processes.

Oxygen is returned to the atmosphere for use in metabolism by photosynthetic green plants and photosynthetic microorganisms such as cyanobacteria. During the process of photosynthesis, these organisms liberate oxygen from water and release it to the atmosphere. The oxygen is then available to heterotrophic organisms for use in their metabolism.

INTERACTIONS AMONG SOIL MICROORGANISMS

Soil is the largest terrestrial ecosystem where a wide variety of relationships exists between different types of soil organisms. The associations existing between different soil microorganisms, whether of a symbiotic or antagonistic nature, influence the activities of microorganisms in the soil. Microflora composition of any habitat is governed by the biological equilibrium created by the associations and interactions of all individuals found in the community. In soil, many microorganisms live in close proximity and interact among them-selves in a different ways. Some of the interactions or associations are mutually beneficial, or mutually detrimental or neutral.

The various types of possible interactions/associations occurring among the microorganisms in soil can be:

- Beneficial
  - Mutualism
  - Commensalisms and
  - Proto-cooperation or
- Detrimental/ harmful -
  - Amensalism,
  - Antagonism,
  - Competition

- Parasitism and
- Predation

BENEFICIAL ASSOCIATION/INTERACTIONS:

- *Mutualism (Symbiosis):* It is a relationship or a type of symbiosis in which both the interacting organisms/partners are benefited from each other. The way/manner in which benefit is derived depends on the type of interactions. When the benefit is in the term of exchange of nutrients, then the relationship is termed as "syntrophism" (Greek meaning: Syn -mutual and trophe = nourishment), for example, Lichen (association of algae or BGA with fungus) in which algae benefits by protection afforded to it by the fungal hyphae from environmental stresses, while the fungus obtain and use CO_2 released by the algae during photosynthesis. Where the blue green algae are the partners in the lichen association, the heterotrophs (Fungus), benefit from the fixed nitrogen by the blue green algae.

  Microorganisms may also form mutualistic relationships with plants, for example nitrogen fixing bacteria *i.e.* Rhizobium growing in the roots of legumes. In this Rhizobium-legume association, Rhizobium bacteria are benefited by protection from the environmental stresses while in turn plant is benefited by getting readily available nitrate nitrogen released by the bacterial partner.

  The Anabaena-Azolla is an association between the water fern Azolla and the cyanobacterium Amabaena. This association is of great importance in paddy fields, where nitrogen is frequently a limiting nutrient.

  An actinorrhizal symbiosis of actinomycetes, Frankia with the roots of Alnus and Casurina (non-legumes) is common in temperate forest ecosystem for soil nitrogen economy. Another type of symbiotic association which exists between the roots of higher plants and fungus is Mycorrhiza. In this association fungus gets essential organic nutrients and protection form roots of the plants and allows them to multiply and in turn plants uptake phosphorus, nitrogen and other inorganic nutrients made available by the fungus.
- *Commensalisms:* In this association one organism/partner in association is benefited by other partner without affecting it. For example, many fungi can degrade cellulose to glucose, which is utilized by many bacteria. Lignin which is major constituent of woody plants and is usually resistant to degradation by most of the microorganisms but in forest soils, lignin is readily degraded by a group of Basidiomycetous fungi and the degraded products are used by several other fungi and bacteria which can not utilize

lignin directly. This type of association is also found in organic matter decomposition process.

- *Proto-cooperation:* It is mutually beneficial association between two species/ partners. Unlike symbiosis, proto-cooperation is not obligatory for their existence or performance of a particular activity. In this type of association one organism favour its associate by removing toxic substances from the habitat and simultaneously obtain carbon products made by the another associate/partner. Nutritional proto-cooperation between bacteria and fungi has been reported for various vitamins, amino and purines in terrestrial ecosystem and are very useful in agriculture. Proto-cooperative associations found beneficial in agriculture are:
  - Synergism between VAM fungus-legume plants andRhizobium in which nitrogen fixation and phosphorus availability/ uptake is much higher resulting in higher crop yields and improved soil fertility,
  - Synergism between PSM-legume plants and Rhizobium and
  - Synergism between plant roots and PGPR in rhizosphere where rhizobacteria restrict the growth of phytopathogens on plant roots and secrets growth promoting substances.

Detrimental (Harmful) Associations/Interactions:

- *Antagonism:* It is the relationship in which one species of an organism is inhibited or adversely affected by another species in the same environment. In such antagonism, one organism may directly or indirectly inhibit the activities of the other. Antagonistic relations are most common in nature and are also important for the production of antibiotics. The phenomenon of antagonism may be categorized into three *i.e.* antibiosis, competition and exploitation.

  In the process of antibiosis, the antibiotics or metabolites produced by one organism inhibits another organism. An antibiotic is a microbial inhibitor of biological origin. Innumerable examples of antibiosis are found in soil. For example, Bacillus Species from soil produces an antifungal agent which inhibits growth of several soil fungi. Several species ofStreptomyces from soil produces antibacterial and antifungal antibiotics. Most of the commercial antibiotics such as streptomycin, chloramphenicol, Terramycin and cyclohexamide have been produced from the mass culture of Streptomyces. Thus, species of Streptomyces are the largest group of antibiotic producer's in soil. Another example of antibiosis is inhibition of Verticillium by Trichoderma, inhibition of Rhizoctonia by a bacterium Bacillus subtilis, inhibition of soil fungus Aspergillus terreus by a bacteriumStaphylococcus aureus.

- *Ammensalism:* In this interaction/association one partner suppress the growth of other partner by producing toxins like antibiotics and harmful gases like ethylene, HCN, Nitrite etc.
- *Competition:* As soil, is inhabited by many different species of microorganisms, there exists an active competition among them for available nutrients and space. The limiting substrate may result in favouring one species over another. Thus, competition can be defined as "the injurious effect of one organism on another because of the removal of some resource of the environment". This phenomenon can result in major fluctuations in the composition of the microbial population in the soil.

  For example, chlamydospores of Fusarium, Oospores of Aphanomyces and conidia ofVerticillium dahlae require exogenous nutrients to germinate in soil. But other fungi and soil bacteria deplete these critical nutrients required for spore germination and thereby hinder the spore germination resulting into the decrease in population. Competition for free space has been, reported to suppress the fungal population by soil bacteria. Therefore, organisms with inherent ability to grow fast are better competitors.
- *Parasitism:* It is an association, in which one organism lives in or on the body of another. The parasite is dependent upon the host and lives in intimate physical contact and forms metabolic association with the host. So this is a host -parasite relationship in which one (parasite) is benefited while other (host) is adversely affected, although not necessarily killed. Parasitism is widely spread in soil communities, for example, bacteriophages (viruses which attack bacteria) are strict intracellular parasites Chytrid fungi, which parasitize algae, as well as other fungi and plants; there are many strains of fungi which are parasitic on algae, plants, animals parasitized by different organisms, earthworms are parasitized by fungi, bacteria, viruses etc.
- *Predation:* Predation is an association/ exploitation in which predator organism directly feed on and kills the pray organism. It is one of the most dramatic inter relationship among microorganisms in nature, for example, the nematophagous fungi are the best examples of predatory soil fungi. Species of Arthrobotrytis and Dactylella are known as nematode trapping fungi. Other examples of microbial predators are the protozoa and slime mold fungi which feed on the bacteria and reduce their population. The bacteriophages may also be considered as predators of bacteria.

MICROBIAL INTERACTIONS AND BIOCONTROL IN THE RHIZOSPHERE

As seeds germinate and roots grow through the soil the loss of organic material provides the driving force for the development of active microbial populations around the root, known as the rhizosphere effect. Although the stimulation in microbial activity is a general phenomenon largely involving saprotrophs, specific groups of symbionts may be selectively enhanced. For example, mutualistic biotrophic symbioses may develop between Rhizobia and legumes, and mycorrhizal fungi may interact with their plant hosts.

However, antagonistic symbioses between pathogens and roots can also form resulting in disease. The microbial interactions taking place in the spermosphere and rhizosphere associated with disease development and especially biocontrol of these diseases form the background of this review.

Interest in biological control has increased recently fuelled by public concerns over the use of chemicals in the environment in general, and the need to find alternatives to the use of chemicals for disease control. The key to achieving successful, reproducible biological control is the gradual appreciation that knowledge of the ecological interactions taking place in soil and root environments is required to predict the conditions under which biocontrol can be achieved and, indeed, may be part of the reason why more biocontrol agents are reaching the market-place.

This type of work requires a study not only of any potential biocontrol agent per se but also its interactions with the crop, the natural resident microbiota and the environment as well.

In this regard, it is well known that some soils are naturally suppressive to some soil-borne plant pathogens such as Fusarium oxysporum Schlect.: Fr. Emend. Snyder and Hansen, Gaeumannomyces graminis (Sacc.) v. Arx and Oliver, Pythium and Phytophthora species and this suppression relates to both physicochemical and microbiological features of the soil (Whipps, 1997a). Importantly, a soil that is suppressive to one pathogen is not necessarily suppressive to another, and so specificity in the soil–plant–microbe interactions for disease suppression exists.

Modern methods for analysing microbial community structures may prove particularly valuable to help define the key organisms or groups of organisms responsible for such natural suppression as well as for monitoring the spread and impact of introduction of specific biocontrol agents or other management practices on natural microbial populations. Significantly, disease suppression can also be achieved by manipulation of the physicochemical and microbiological environment through management practices such as the use of soil amendments, crop rotations, use of fumigants or soil solarization. However, at present, greatest interest resides with the development and application of specific biocontrol agents for the control of diseases on seeds and roots and the interaction of these with pathogens and hosts, and will form the focus of this chapter.

There have been numerous reviews in recent years on this topic and so only selected recent examples, illustrating key features of biocontrol on seeds and roots, particularly the different modes of action, will be discussed whenever possible. Modes of action include: inhibition of the pathogen by antimicrobial compounds (antibiosis); competition for iron through production of siderophores; competition for colonization sites and nutrients supplied by seeds and roots; induction of plant resistance mechanisms; inactivation of pathogen germination factors present in seed or root exudates; degradation of pathogenicity factors of the pathogen such as toxins; parasitism that may involve production of extracellular cell wall-degrading enzymes, for example, chitinase and ß-1,3 glucanase that can lyse pathogen cell walls.

None of the mechanisms are necessarily mutually exclusive and frequently several modes of action are exhibited by a single biocontrol agent. Indeed, for some biocontrol agents, different mechanisms or combinations of mechanisms may be involved in the suppression of different plant diseases.

BACTERIA–BACTERIAL PATHOGEN INTERACTIONS

In the last few years there have been relatively few studies of bacteria applied to seeds and roots for the purpose of controlling bacterial diseases. One example, is the application of non-pathogenic strains ofStreptomyces to control scab of potato (Solanum tuberosum L.) caused byStreptomyces scabies (Thaxter) Waksman and Henrici. Here biocontrol may operate through antibiosis or competition for space or nutrients in the rhizosphere.

In contrast, Pseudomonas fluorescens (Trevisan) Migula F113 was shown to control the soft rot potato pathogen Erwinia carotovora subsp. atroseptica (van Hall) Dye by production of the antibiotic 2,4-diacetylphloroglucinol (DAPG) and, through use of co-inoculation experiments with mutants lacking DAPG production, that competition was not a feature of biocontrol in this system (Cronin et al., 1997).

Some evidence was also obtained that siderophore production byP. fluorescens F113 may play a role in biocontrol of potato soft rot under iron-limiting conditions, but DAPG appears to be the major biocontrol determinant. Pseudomonas species may also control crown gall disease in many dicotyledonous plants caused by Agrobacterium tumefaciens (Smith and Townsend) Conn (Khmel et al., 1998).

However, the classic, and still commercially successful, bacterial-based biocontrol system is the use of non-pathogenic Agrobacterium strains to control Agrobacterium tumefaciens. Long-term molecular and ecological studies of this control system have identified how the biocontrol works and have also allowed potential problems associated with its use in the field to be overcome.

The most widely used non-pathogenic Agrobacterium strain K84 produces a highly specific antibiotic agrocin 84, which is encoded by plasmid pAgK84. Inundative inoculation of Agrobacterium strain K84 to roots by dipping in cell suspensions prior to exposure to the pathogen effectively controls those strains of pathogen susceptible to agrocin 84. However, because there is a risk that plasmid pAgK84 could be transferred to pathogenic strains and reduce effectiveness of control, a transfer deletion mutant of K84, K1026 has been constructed (Jones et al., 1988).

Strain K1026 is as efficient as K84 in biocontrol of both strains susceptible to agrocin 84 and those resistant to agrocin 84 and so, clearly, production of agrocin 84 is not the only mechanism of biocontrol. Production of other antibiotics such as agrocin 434 or ALS 84 may play a part, but the ability to survive and compete on roots may also be important. Studies where pathogenic cells were co-inoculated with K84 or K1026 resulted in survival of the pathogen on roots up to 8 months later, although no symptoms were present, providing evidence that the non-pathogenic strains prevented disease expression rather than killing pathogen cells directly.

BACTERIA–FUNGAL PATHOGEN INTERACTIONS

The volume of literature in this area continues to increase at a rapid rate, stimulated by the increasing ease with which molecular techniques can be applied to answer questions concerning distribution, and occurrence and relative importance of specific modes of action. Some examples of the different types of bacteria-fungal pathogen interactions examined in the spermosphere and rhizosphere in just the last three years.

Although a range of different bacterial genera and species have been studied, the overwhelming number of papers have involved the use of Pseudomonas species. Clearly, Pseudomonas species must have activity but it begs the question as to the features that make this genus so effective and the choice of so many workers. Pseudomonads are characteristically fast growing, easy to culture and manipulate genetically in the laboratory, and are able to utilize a range of easily metabolizable organic compounds, making them amenable to experimentation. But, in addition, they are common rhizosphere organisms and must be adapted to life in the rhizosphere to a large extent.

Having appropriate ecological rhizosphere competence may be a key feature for reproducible biological control activity in the spermosphere and rhizosphere. This criterion is already widely appreciated for many fungal biocontrol agents. A few specific examples of the modes of action involved with bacterial biocontrol of fungal pathogens in the rhizosphere and spermosphere are given below.

Antibiosis

There are numerous reports of the production of antifungal metabolites (excluding metal chelators and enzymes) produced by bacteria in vitrothat may also have activity in vivo. These include ammonia, butyrolactones, 2,4-diacetylphloroglucinol (Ph1), HCN, kanosamine, Oligomycin A, Oomycin A, phenazine-1-carboxylic acid (PCA), pyoluterin (Plt), pyrrolnitrin (Pln), viscosinamide, xanthobaccin, and zwittermycin A as well as several other uncharacterized moieties. To demonstrate a role for antibiotics in biocontrol, mutants lacking production of antibiotics or over-producing mutants have been used.

Alternatively, the use of reporter genes or probes to demonstrate production of antibiotics in the rhizosphere is becoming more commonplace. Indeed, isolation and characterization of genes or gene clusters responsible for antibiotic production has now been achieved. Significantly, both Phl and PCA have been isolated from the rhizosphere of wheat following introduction of biocontrol strains of Pseudomonas, finally confirming that such antibiotics are produced in vivo. Further, Ph1 production in the rhizosphere of wheat was strongly related to the density of the bacterial population present and the ability to colonize roots. PCA from Pseudomonas aureofaciens Kluyver Tx-1 has even been used as a direct field treatment for the control of dollar spot (Sclerotinia homeocarpa F. T. Bennett) on creeping bentgrass (Agrostis palustrisHudson).

Antibiotic production by bacteria, particularly pseudomonads, seems to be closely regulated by a two-component system involving an environmental sensor (presumably a membrane protein) and a cytoplasmic response factor. Mutation in either gene has similar multiple effects on antibiotic production. For example, P. fluorescens Pf-5 with a mutation in the apdA sensor gene lost the ability to produce HCN, Plt and Pln and P. fluorescens CHA0 with a defect in the gacA response gene lost the ability to produce Phl, Plt and HCN as well as protease and phospholipase C. However, the environmental signals that control the two-component system are unknown. Interestingly, the gacA gene is required for biocontrol activity in P. fluorescens CHA0 in the rhizosphere of dicotyledous plants, but not in those of the Gramineae, although the mechanisms are unclear.

Other two-component signalling phenomena may also be involved in PCA production by pseudomonads on roots. Pseudomonas aureofaciens 30-84 is a biocontrol agent of take-all disease of wheat (Triticium aestivum L.) caused by Gaeumannomyces graminis var. tritici Walker and operates through inhibiting growth of the pathogen by production of PCA. In this system, pathogen growth on the root increases root exudation and this results in an increase in the population of P. aureofaciens 30-84 and other bacteria in the infection zone.

Consequently, there is an increase in the level of the signal molecule N-acyl-L-homoserine lactone (HSL) produced at low levels constitutively by the phzI gene, in the rhizosphere which is sufficient to switch on the PCA synthesis pathway in P. aureofaciens 30-84 controlled by the phzR gene. The resulting PCA production inhibits further growth of the pathogen. This explains why P. aureofaciens 30-84 does not reduce the number of infection sites on the roots, but inhibits secondary growth of the pathogen. Significantly, HSL from other members of the rhizosphere microbial community can contribute to PCA production in P. aureofaciens30-84 raising the question of the significance of interpopulation signalling on biocontrol and perhaps the enhanced performance of certain strains of bacteria when introduced with mixtures of other bacterial biocontrol strains (Pierson and Weller, 1994).

In addition, antibiotic production in Pseudomonas spp. may be further controlled by the activity of housekeeping sigma factors encoded by rpoS or rpoD genes, illustrating the complexity of these regulatory systems.

Interestingly, signalling between pathogenic fungi and potential biocontrol bacteria has also been detected. In one case, trehalose derived from Pythium debaryanum Hesse up-regulated genes in its biocontrol strain Pseudomonas fluorescens ATCC 17400 (Gaballa et al., 1997) and yet in another example Pythium ultimum Trow caused a down-regulation of five gene clusters of P. fluorescens F113 which provides biocontrol of this pathogen in the rhizosphere of sugar beet (Beta vulgaris L.) (Fedi et al., 1997). These findings may be of considerable significance for bacterial–fungal interactions in general and has major implications for the control of gene expression in complex microbial communities.

Competition for Iron

Although competition between bacteria and fungal plant pathogens for space or nutrients has been known to exist as a biocontrol mechanism for many years the greatest interest recently has involved competition for iron. Under iron-limiting conditions, bacteria produce a range of iron chelating compounds or siderophores which have a very high affinity for ferric iron. These bacterial iron chelators are thought to sequester the limited supply of iron available in the rhizosphere making it unavailable to pathogenic fungi, thereby restricting their growth.

Recent studies have clearly shown that the iron nutrition of the plant influences the rhizosphere microbial community structure. Iron competition in pseudomonads has been intensively studied and the role of the pyoverdine siderophore produced by manyPseudomonas species has been clearly demonstrated in the control ofPythium and Fusarium species, either by comparing the effects of purified pyoverdine with synthetic iron chelators or through the use of pyoverdine minus mutants.

Pseudomonads also produce two other siderophores, pyochelin and its precursor salicylic acid, and pyochelin is thought to contribute to the protection of tomato plants from Pythium by Pseudomonas aeruginosa (Schroeter) Migula 7NSK2. However, siderophores are not always implicated in disease control by pseudomonads. The dynamics of iron competition in the rhizosphere are often complex. For example, some siderophores can only be used by the bacteria that produce them, whereas others can be used by many different bacteria. Different environmental factors can also influence the quantity of siderophores produced. There is also the further complication that pyoverdine and salicylate may act as elicitors for inducing systemic resistance against pathogens in some plants.

PARASITISM AND PRODUCTION OF EXTRACELLULAR ENZYMES

The ability of bacteria, especially actinomycetes, to parasitize and degrade spores of fungal plant pathogens is well established. Assuming that nutrients pass from the plant pathogen to bacteria, and that fungal growth is inhibited, the spectrum of parasitism could range from simple attachment of cells to hyphae, as with the Enterobacter cloacae (Jordan) Hormaeche and Edwards–Pythium ultimum interaction to complete lysis and degradation of hyphae as found with the Arthrobacter–Pythium debaryanum interaction. If fungal cells are lysed and cell walls are degraded then it is generally assumed that cell wall-degrading enzymes produced by the bacteria are responsible, even though antibiotics may be produced at the same time.

Considerable effort has gone into identifying cell wall-degrading enzymes produced by biocontrol strains of bacteria even though relatively little direct evidence for their presence and activity in the rhizosphere has been obtained. For example, biocontrol of Phytophthora cinnamomi Rands root rot of Banksia grandis Willd. was obtained using a cellulase-producing isolate of Micromonospora carbonacea Luedemann and Brodsky and control of Phytophthora fragariaevar. rubi Hickman causing raspberry root rot was suppressed by the application of actinomycete isolates that were selected for the production of β-1,3-, β-1,4- and β-1,6-glucanases.

Chitinolytic enzymes produced by both Bacillus cereus Frankland and Pantoea(Enterobacter) agglomerans (Beijerinck) Gavini et al. also appear to be involved in biocontrol of Rhizoctonia solani Kühn. Tn5 mutants of E. agglomerans (Beijerinck) Gavini et al. deficient in chitinolytic activity were unable to protect cotton (Gossypium barbardense L.) and expression of the chiA gene for endochitinase in Escherichia coli (Migula) Castellani and Chalmers allowed the transformed strain to inhibit R. solani on cotton seedlings.

Similar techniques involving Tn5 insertion mutants and subsequent complementation demonstrated that biocontrol of Pythium ultimum in the rhizosphere of sugar beet by Stenotrophomonas maltophila (Hugh) Palleroni and Bradbury W81 was due to the production of extracellular protease.

Induced Resistance

Perhaps the greatest growth area in biocontrol in the last few years has been concerned with induced resistance defined as 'the process of active resistance dependent on the host plant's physical or chemical barriers, activated by biotic or abiotic agents (inducing agents)'. This has come about through the synergistic interaction of microbiologists, plant pathologists and plant scientists armed with an appropriate battery of molecular tools. The effect had previously often been overlooked through inadequate techniques or controls as well as the biocontrol agent exhibiting other modes of action at the same time.

Most work has focused on the systemic resistance induced by non-pathogenic rhizosphere-colonizing Bacillus and Pseudomonas species in systems where the inducing bacteria and the challenging pathogen remained spatially separate for the duration of the experiment, and no direct interaction between the bacteria and pathogen was possible. Such split root or spatial root inoculation experiments were used to demonstrate the phenomenon in radish (Raphanus sativus L.) and Arabidopsis against Fusarium oxysporum and in cucumber (Cucumis sativus L.) against Pythium aphanidermatum (Edson) Fitzp. Various combinations of timing and position have indicated that induced resistance also occurs in carnation (Dianthus caryophyllus L.), tobacco (Nicotiana tabacum L.) and tomato (Lycopersicon esculentum Mill.).

Bacteria differ in ability to induce resistance, with some being active on some plant species and not others; variation in inducibility also exists within plant species. The full range of inducing moieties produced by bacteria is probably not yet known, but lipopolysaccharides and siderophores are clearly indicated.

The definition of induced resistance suggested by Kloepper et al. covered both biotic and abiotic inducers. Although the phenotypic effects of root inoculation with bacteria may be similar to treatment with abiotic agents or micro-organisms that cause localized damage, the biochemical and mechanistic changes appear to be subtly different. This has resulted in the term induced systemic resistance (ISR) for bacterially-induced resistance and systemic acquired resistance for the other forms.

The major differences are that pathogenesis-related (PR) proteins such as chitinases, β-1,3-glucanases, proteinase inhibitors and one or two other rarer types, are not universally associated with bacterially induced resistance and salicylic acid (a known inducer of SAR) is not always involved in expression of ISR, but this is dependent on bacterial strain and host plant involved.

Ethylene responsiveness may also be required at the site of inoculation of the inducing bacteria for ISR to occur:

Changes that have been observed in plant roots exhibiting ISR include:

- Strengthening of epidermal and cortical cell walls and deposition of newly formed barriers beyond infection sites including callose, lignin and phenolics;
- Increased levels of enzymes such as chitinase, peroxidase, polyphenol oxidase, and phenylalanine ammonia lyase;
- Enhanced phytoalexin production;
- Enhanced expression of stress-related genes.

However, not all of these biochemical changes are found in all bacterial–plant combinations. Similarly, the ability of bacteria to colonize the internal tissue of the roots has been considered to be an important feature in many of the bacterial–root interactions involving ISR, but is not a constant feature of them all.

SYMBIOSES WITH SOIL-BASED MICROORGANISMS

Nitrogen and phosphorus are among the elements considered most limiting to plant growth and productivity because they are often present in small quantities locally or are present in a form that cannot be used by the plant.

As a result, the evolution of many plant species has included the development of mutually beneficial symbiotic relationships with soil-borne microorganisms. In these relationships, both the host plant and the microorganism symbiont derive valuable resources that they need for their own productivity and survival as a result of the association.

Nitrogen Fixation. Despite the fact that nitrogen is the most abundant gaseous element in the atmosphere, plants are unable to utilize the element in this form (N_2) and may experience nitrogen deficiency in some soils that have low nitrogen content. Since nitrogen is a primary component of both proteins and nucleic acids, nitrogen deficiency imposes significant limitations to plant productivity.

In an agricultural setting, nitrogen deficiency can be combated by the addition of nitrogen-rich fertilizers to increase the availability of nutrients and thereby increase crop yield. However, this can be a dangerous practice since excess nutrients generally end up in ground water, leading to eutrophication and subsequent oxygen deprivation of connected aquatic ecosystems.

Plants are able to directly acquire nitrate and ammonium from the soil. However, when these nitrogen sources are not available, certain species of plants from the family Fabaceae (legumes) initiate symbiotic relationships with a group of nitrogen fixing bacteria called Rhizobia.

These interactions are relatively specific and require that the host plant and the microbe recognize each other using chemical signals. The interaction begins when the plant releases compounds called flavanoids into the soil that attract the bacteria to the root. In response, the bacteria release compounds called Nod Factors (NF) that cause local changes in the structure of the root and root hairs. Specifically, the root hair curls sharply to envelop the bacteria in a small pocket. The plant cell wall is broken down and the plant cell membrane invaginates and forms a tunnel called an infection thread that grows to the cells of the root cortex.

The bacteria become wrapped in a plant derived membrane as they differentiate into structures called bacteroids. These structures are allowed to enter the cytoplasm of cortical cells where they convert atmospheric nitrogen to ammonia, a form that can be used by the plants. In return, the bacteroids receive photosynthetically derived carbohydrates to use for energy production (reviewed by Limpens and Bisseling, 2003; Ferguson et al. 2010).

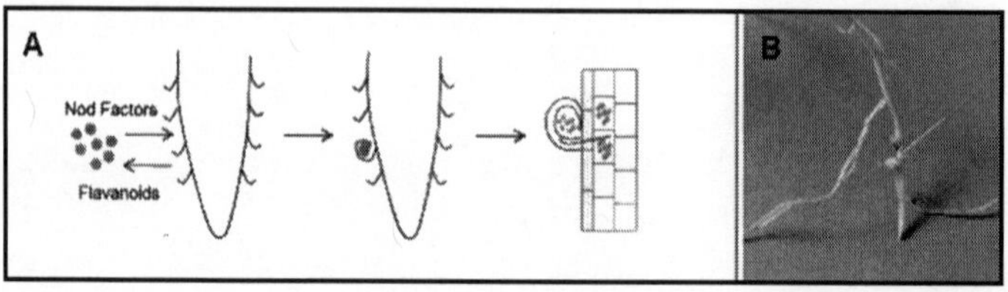

Fig. Nodulation of Legumes.

A. Process of root cell colonization by rhizobacteria. B. Nodule formed by nitrogen fixing bacteria on a root of a pea plant (genus Pisum).

MYCORRHIZAL INTERACTIONS WITH PLANTS

In addition to symbiotic relationships with bacteria, plants can participate in symbiotic associations with fungal organisms as well. Current estimates of the frequency of plant-mycorrhizal associations suggest that around 80 per cent of all plants establish some type of mycorrhizal symbiosis, and many studies indicate that these relationships are millions of years old.

There are several classes of mycorrhiza, differing in structural morphology, the method of colonizing plant tissue, and the host plants colonized.

However, there are two main classes that are generally regarded as the most common and therefore, the most ecologically significant. The endomycorrhizae are those fungi that establish associations with host plants by penetrating the cell wall of cortical cells in the plant roots. By contrast, ectomycorrizae develop a vast hyphae network between cortical cells but do not actually penetrate the cells.

The most common endomycorrhizal interaction occurs between arbuscular mycorrhizal fungi (AMF; also called Vesicular-Arbuscular Mycorrhiza or VAM) and a variety of species of grasses, herbs, trees and shrubs. When phosphate is available in the soil, plants are able to acquire it directly via root phosphate transporters. However, under low phosphate conditions, plants become reliant on interactions with mycorrhizal fungi for phosphorus acquisition. Mycorrhizal spores present in the soil are germinated by compounds released from the plant.

Hyphae extend from the germinating spore and penetrate the epidermis of the plant root. Inside the root, the hyphae branch and penetrate cortical cells, where highly branched structures called arbuscules develop. Externally, hyphae extend into the soil beyond the area accessible to the root.

This kind of symbiosis facilitates plant phosphorus uptake from the soil by increasing the root's absorptive surface area. Since plants take up phosphorus at a much higher rate than phosphorus diffuses into the soil surrounding the root, a phosphorus depletion zone is quickly established, limiting uptake of phosphorus by the plant. However, AMF hyphae form a bridge between the internal root environment and the area beyond this depletion zone enabling the plant to acquire significantly more phosphorus through its symbiotic partner than it could on its own.

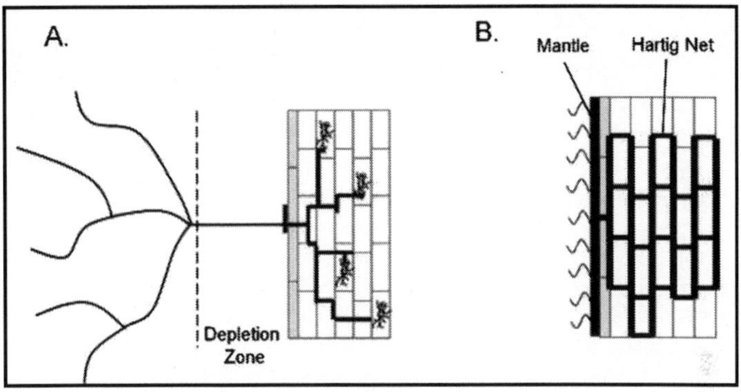

Fig. Plant-mycorrhizal Fungus Interactions.

A. Diagram of arbuscular mycorrhizae colonization of a plant root showing the extension of hyphae beyond the phosphorus depletion zone and the presence of arbuscules in cells of the root cortex. B. Diagram of Ectomycorrhizal fungi showing growth of hyphae around cortical cells, a mantle sheath on the outside of the root, and hyphae that extend into soil around the root.

Ectomycorrhizal fungi (EcM) form associations with many tree species (birch, oak, spruce, pine, fir), making these microorganisms an important component of the biodiversity found in forest ecosystems, particularly temperate and boreal forests in the northern hemisphere.

Although the molecular mechanisms underlying initiation of EcM-plant interactions are not well defined, it is clear that signaling molecules are released by both symbiotic partners, resulting in fungal hyphae growth towards compatible plant roots.

The EcM forms an extensive hyphal sheath, or mantle, around the root cap and a hartig net of hyphae that extends into the root itself and surrounds the cells within the root cortex. Hyphae from the mantle extend into the surrounding soil, providing availability and storage of nutrients like phosphorus and nitrogen for the plant. This is particularly important for trees growing in forest soils because a large portion of the nutrients available are located in the litter layer and are not accessible to the roots of these trees.

The EcM produce enzymes that can digest this organic material and also have the ability to mobilize the nutrients to the hartig net, making them available to the plant. In this way, the interaction between trees and EcM is not only beneficial, but in many cases, is essential (reviewed by Nehls et al., 2008). As is the case with bacterial symbioses, the benefit to the fungal partners in these relationships involves transfer of photosynthetic carbohydrates from the plant to the microorganism for metabolism and energy production. However, the nutritional advantage for plants of such symbiotic relationships far outweighs the energetic cost imposed by hosting microorganisms like rhizobacteria, AMF, and EcM.

METHODS OF SOIL MICROBIAL COMMUNITY

Utilizing a common garden of known genotypes planted in 1991, we quantified the soil microbial community beneath three to five replicate clones each of multiple genotypes of Populus fremontii (four genotypes), P. angustifolia (five genotypes) and their natural F1 (five genotypes) and backcross hybrids (six genotypes) for a total of 70 trees. Tree genotypes were previously collected and replicated as stem cuttings from natural populations along a 105 km transect along the Weber River, northern UT, USA and planted randomly in a one hectare common garden in Ogden UT, on 5 meter centers.

Randomization of replicate clones within the garden minimized site bias and maximized our ability to detect genetic effects. Soils at the garden were typical of low elevation alluvial riparian sites and were characterized as coarse-loamy, mixed, mesic Oxyaquic Haploxerolls. Due to 13 years of Populus litter accumulation and shading, the understory is sparse immediately adjacent to the tree trunk but includes Bromus tectorum, Ambrosia spp., Tragopogon dubious, Poa bulbusa, Lactuca serriola; the understory community showed no pattern of abundance relative to the tree genotypes or cross types (Kovacs et al. pers. comm.).

Tree genotypes and specific cross type status (*i.e.,* pure species, F1 or backcross hybrids) were previously determined with restriction fragment length polymorphism (RFLP) analyses (Martinsen et al. 2001). While genetic by environment interactions (G X E) are clearly important to microbial dynamics, common garden approaches to elucidate the role of the genetic vs. environmental components are valuable to help tease apart the genetic interactions that may tie plants to their associated communities in ways that are only beginning to be appreciated.

MICROBIAL ANALYSES

In October 2004, when the trees were 13 years old (and 10-15 m tall), we sampled bulk, mineral soils beneath each tree using polycarbonate soil cores (4.8 cm X 15 cm).

We chose this season as previous research in this system indicated that fall was an optimal time to quantify microbial activity due to increased rainfall and pulses of carbon from litterfall (as demonstrated by highest rates of net nitrogen mineralization, relative to other seasons in the year; data from Schweitzer et al. 2004).

To minimize disturbance beneath each tree in this long-term common garden, soil core placement was standardized; a 15 cm deep soil core was placed and removed on the north side of each tree within 0.25 m of the trunk. The soils were kept cool, sieved (< 2 mm) to remove coarse fragments and roots, and a sub-sample was immediately freeze-dried. From these samples, we assessed microbial community composition with PLFA biomarkers using the method of White and Ringleberg (1998).

The freeze-dried soil was extracted with a phosphate-buffered chloroform-methanol solvent (Bligh and Dwyer 1959) and separated into functional classes of signature fatty acids after methylation of the polar lipids and analysis by gas chromatography (White et al. 1979; Agilent Technologies GC-Mass Spectrometer [6890N GC/ 5973N MSD] Palo Alto, CA, USA). The PLFAs are unique to major taxonomic groups. While this is a coarse measure of microbial community composition, recent evidence suggests it is just as effective in differentiating treatment patterns as functional analyses (*i.e.*, community-level physiological profiling) and molecular techniques (Ramsey et al. 2006).

We conservatively used common PLFA between 14 and 18 carbon (C) atoms long to include in the analysis as those are PLFA exclusively identified as microbial. Total PLFA was determined as the sum of all PLFA biomarker concentrations. PLFA compounds identified as general bacterial and fungal biomarkers were used to calculate the ratio of bacterial to fungal PLFA concentration in the soils.

We also quantified microbial biomass N using the chloroform fumigation extraction method (Haubensak et al. 2002). A sub-sample of field moist soil (ca. 25 g) was extracted with 50 ml 0.5 M K2SO4, shaken, gravity-filtered with Whatman filter paper no. 1 (leached with deionized water and K2SO4) and stored in a freezer until chemical analysis.

Another 25 g field moist sub-sample was exposed to chloroform for five days in a glass vacuum desiccator with 30 ml ethanol-free chloroform. After the fumigation period the samples were extracted with 50 ml of 0.5 M K2SO4 in the same manner as above.

Total microbial biomass N was determined on the thawed extracts with a micro-Kjeldahl digestion followed by colourimetric analysis using the salicylate method on a Lachat AE auto-analyzer (Lachat Industries, Inc., Loveland, CO, USA).

Microbial biomass N was determined by subtracting the un-fumigated sample from the fumigated sample.

A third sub-sample of field moist soil was oven dried (48 h at 105° C) to determine soil moisture content. All PLFA and microbial biomass values are expressed on an oven-dry mass basis.

Statistical Analyses

To detect differences in microbial biomass and microbial communities (based on the NMDS scores) among tree cross types, we performed a line-cross analysis of the size of the microbial biomass and the composition of the microbial community across the four cross types (Populus fremontii, P. angustifolia, and their natural F1 and backcross hybrids). A line-cross analysis estimates the additive (*i.e.,* parent to offspring genetic effects), dominant (*i.e.,* non-additive, one allele effects), and epistatic (*i.e.,* effects of non-additive novel combinations of genes) components of phenotypic variance associated with plant cross type and microbial dynamics across the hybridizing complex (Lynch and Walsh 1998). This allowed us to make inferences on the relative contribution of plant genotype versus hybrid cross type to microbial dynamics as well as how genetic introgression may affect microbial community dynamics.

We assessed plant genetic effects on microbial communities across the hybridizing complex following the methods of Shuster et al. (2006). We calculated a dissimilarity matrix of the PLFA from soils collected beneath individual tree genotypes using the Bray-Curtis dissimilarity coefficient (Legendre and Legendre 1998).

The major directions of microbial compositional change were summarized and quantified using non-metric multi-dimensional scaling (NMDS) with a one-dimensional solution (Minchin 1987). We extracted the NMDS score which represents an estimate of the combined PLFA richness (*i.e.,* functional classes) and abundance (based on nmol PLFA/g of dry soil) for each replicated clone of each tree genotype.

To analyze differences in microbial biomass and microbial community composition among genotypes within each cross type, we performed a univariate ANOVA of microbial data within each cross type using JMP 5.1 (SAS Institute Cary, NC, USA).

The data showed no evidence for violation of assumptions of normality and homogeneity of variance. From the variance component estimates from the ANOVA analyses, we calculated the broad-sense heritability (*i.e.,* clonal repeatability) of microbial biomass (H^2B ± 95 per cent confidence limits) and microbial community composition (H^2C ± 95 per cent confidence limits) as the among-genotype component of variation in the community phenotype (σ^2_S) divided by the total variance in the community phenotype for all replicate trees (σ^2_{total}), where the total variance in the phenotype was the sum of the among genotype and the error component (σ^2_e Falconer and McKay 1996, Conner and Hartl 2004, Shuster et al. 2006).

To determine if particular PLFA were structuring the microbial community (as opposed to a whole community response), we used Indicator Species Analysis (Dufrene and Legendre 1997, Legendre and Legendre 1998) using PC-ORD 4.02 (MjM Software Design, Gleneden

Beach OR, USA). This analysis calculates an indicator value based on fidelity and relative abundance of species (in this case, PLFA) to each group (in this case plant genotype). Indicator species (those with significant Indicator Values) are defined as those expressing the highest degree of specificity and fidelity to a group or site, independent of abundance, following this basic formula:

$$A_{ij} \times B_{ij} \times 100 = IndVal_{ij} \text{ and } \max [IndVal_{ij}] = IndVal_i$$

A_{ij} is a specificity measure (A_{ij} = Nindividualsij/Nindividualsi) and B_{ij} is a fidelity measure (B_{ij} = $Nsites_{ij}/Nsites_i$; Dufrene and Legendre 1997, Legendre and Legendre 1998, Wimp et al. 2005). A significant indicator species for a particular site or group was determined by a Monte Carlo test based on Bray Curtis distance (Dufrene and Legendre 1997).

MICROBIAL BIOMASS

The line-cross analyses used to estimate the plant genetic contributions to microbial dynamics across the hybridizing complex show that the model incorporating additive genetic effects was not rejected (χ^2=5.98 and χ^2=0.05; df=2, for total PLFA and microbial biomass N, respectively).

However, the additive genetic coefficient was not significant, indicating that variation in microbial biomass within cross types overwhelms the variation among tree cross types (data not shown).

In other words, plant genotype explained an overwhelming proportion of the variance in total PLFA and microbial biomass N relative to the differences between the species and their naturally occurring hybrids.

Consistent with this observation, we found significant differences in microbial biomass N within cross types. Microbial biomass N differs among genotypes within P. angustifolia ($F_{4,14}$ = 4.95, p = 0.026), as well as the F_1 ($F_{4,16}$ = 6.01, p = 0.007) and backcross hybrids ($F_{5,19}$ = 3.67, p = 0.025). While there was variation, there were no significant difference in microbial biomass N among genotypes within P. fremontii ($F_{3,13}$ = 0.28, p = 0.84).

From the variance estimates for microbial biomass N, we found significant broad-sense heritability among genotypes within P. angustifolia (H^2B = 0.61 ± 0.59; [95 per cent confidence intervals]), the F_1 hybrids (H^2B = 0.23 ± 0.09) and the backcross hybrids.

Similarly, for total PLFA (another estimate of microbial biomass), we found significant differences among genotypes in P. angustifolia ($F_{4,14}$ = 15.57, p = 0.0002) and the backcross hybrids ($F_{5,19}$ = 3.25, p = 0.04). There were no differences among genotypes for P. fremontii or the F1 hybrids. The variance estimates for total PLFA suggest that plant genotype explained 78 per cent (H^2B = 0.78 ± 0.032) of the variation within P. angustifolia genotypes and 43 per cent of the variation within backcross hybrids (H^2B = 0.43 ± 0.43, although this is not significant due to overlapping of confidence intervals).

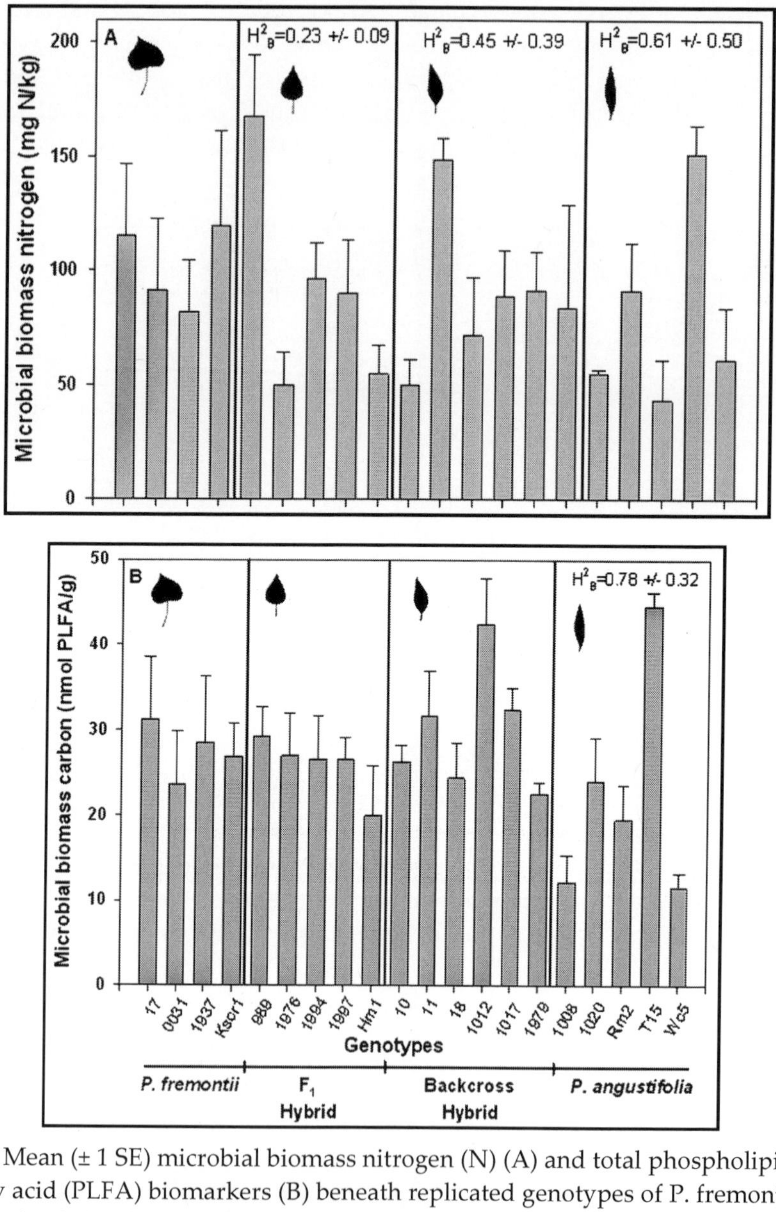

Fig. Mean (± 1 SE) microbial biomass nitrogen (N) (A) and total phospholipids fatty acid (PLFA) biomarkers (B) beneath replicated genotypes of P. fremontii, F1 hybrids, backcross hybrids and Populus angustifolia in a common garden. Broad-sense heritabilities (H^2B ± 95 per cent confidence limits) are listed where significant.

These results indicate that tree genotypes within species (*i.e.*, intraspecific genetic variation) can create environments to which soil microorganisms consistently respond, leading to potential feedbacks in nutrient availability and C storage in soils beneath individual trees.

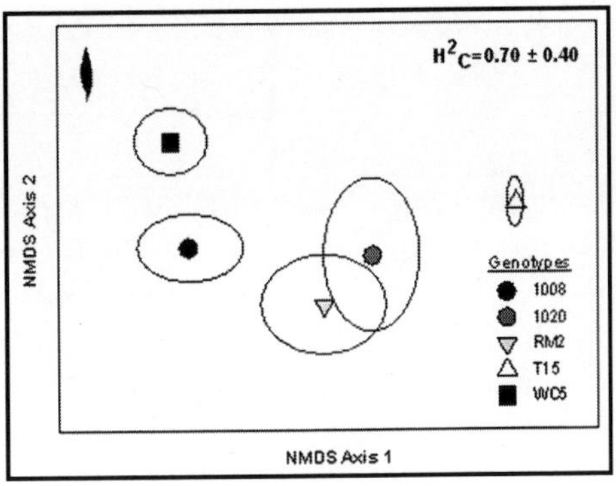

Fig. Mean microbial community composition (*i.e.*, Bray-Curtis dissimilarity) of an individual plant genotype (surrounded by ellipses representing ± 1 SE; n = 3-5 replicates per genotype) underneath replicated tree genotypes of Populus angustifolia in a common garden. As this is a distance measure, the axes are unitless. H^2C (± 95 per cent confidence limits) represents the broad-sense heritability for the microbial community composition.

Microbial Community Composition

A line-cross analysis revealed that additive (χ^2=9.79) and additive plus dominance models (χ^2=8.63) were rejected (df = 2; data not shown), suggesting that variation in microbial community composition within tree cross types was higher than differences among tree cross types. These results also suggest that epistatic genetic interactions across the hybridizing complex likely have a large effect on the microbial community dynamics among the cross types.

When we examined microbial community composition of genotypes within each cross type, we only found significant differences in microbial community composition among P. angustifolia genotypes ($F_{4,14}$ = 8.31, p = 0.0002). Plant genetic factors explained 70 per cent of the variation (*i.e.*, H^2C = 0.70 ± 0.40) in soil microbial communities within P. angustifolia.

Furthermore we found no significant patterns in the types of microorganisms associated with the tree cross types, indicating that our patterns were not driven by specific categories or "blooms" of microorganisms in the soil.

There were no differences within any of the cross types in the ratio of bacterial to fungal PLFA among genotypes (data not shown). Similarly, when we examined the effects of plant genetic factors on individual PLFA biomarkers we only found patterns within backcross hybrids, suggesting that in the pure species and F1 hybrids, no single PLFA was specific to those tree types.

PLFA i15:0, a15:0, 17:0, cy17:0 and 18:2ω6t were significant indicators for backcross hybrids, suggesting strong associations of the soil microorganisms containing these biomarkers with backcross tree genotypes alone. However as there were no overall significant differences in community composition between genotypes in backcross hybrids, the meaning of this result is unclear.

Discussion

After 13 years of plants conditioning the soils beneath them, we found that intraspecific plant genetic variation can affect soil microbial biomass and microbial community composition. These data support the hypothesis that feedbacks from plant genotype may be an important ecological factor affecting microbial dynamics. Overall, we found that plant genotype influenced microbial biomass N beneath replicated genotypes of P. angustifolia, F1 and backcross hybrids (but not P. fremontii genotypes) and demonstrated significant broadsense heritability.

Plant genetic factors explained 61 per cent, 23 per cent and 41 per cent of the variation in microbial biomass N, respectively. Total PLFA, as another measure of microbial biomass, demonstrated significant genotype effects in soils associated with P. angustifolia and backcross hybrid genotypes. Genotype was a significant predictor of soil microbial community composition within genotypes of P. angustifolia, with plant genotype explaining 70 per cent of the variation in community composition.

Among cross types (excepting backcross hybrids) we found no specific PLFA were associated or specific to tree cross type; the bacteria to fungi ratio also did not differ among genotypes in any of the tree cross types. The lack of significant indicators suggests that the observed patterns were not driven by any particular microbial group. In total these results demonstrate that microbial biomass N and sometimes microbial community composition vary by plant genotype, which suggests that plant genotype can select for particular soil microbial characteristics (Kukui 2003).

Genotypic Variation

While the studies to date in this system have focused on the pure and hybrid cross type level (Whitham et al. 2003, 2006), only recently have we begun to investigate the within class variation (intraspecific genetic variation). For example, when plant genotype was examined, genotype explained a greater proportion of the variance in arthropod community composition than species or hybrid cross type (Shuster et al. 2006).

Moreover, the patterns of broad-sense heritability of arthropod community composition followed a similar gradient with the highest H2C occurring in the P. angustifolia and backcross hybrids and the lowest H2C in the P. fremontii (Shuster et al. 2006). The consistency of these two studies suggest that there are similar patterns of genetic variation within the pure and hybrid cross types (*i.e.*, low genetic variation in P. fremontii and high genetic variation in P. angustifolia), which is consistent with past research (Keim et al. 1989).

Results from these two studies indicate that, regardless of the response variable (*i.e.*, arthropod communities – see Shuster et al. 2006, microbial biomass N, and microbial community composition – this study), intraspecific plant genetic variation explains a greater proportion of the variance than do the differences between species and their hybrids. From an evolutionary perspective this should not be unexpected.

Populations of genotypes that are distributed across the landscape face a variety of biotic and abiotic selective forces which results in genotypes with unique suites of traits, genetic architecture, and therefore unique evolutionary dynamics (Thompson 2005). Other studies of genotype-specific responses in many response variables (*e.g.* arthropod communities, trophic-level interactions and litter decomposition) are being demonstrated across plant species and suggest that there may be much genetic variation (for any given trait) within species.

Demonstration of broad-sense heritability, or clonal repeatability, in microbial biomass and community composition (at least in P. angustifolia) suggests that tree genotypes within species (*i.e.*, intraspecific genetic variation) can create environments to which soil microorganisms may respond. However in natural systems, experiments at the genotype level examining plant-microbial interactions have been inconclusive, mostly due to lack of replication.

The studies to date also suggest that belowground components (*i.e.*, roots and/or root exudates) may play a large role in structuring the soil microbial community. For example, Bever et al. (1996) found no effect of plant genotype on arbuscular mycorrhizal communities in a greenhouse study, although the study only utilized one genotype each of four different plant species. Similarly, Madritch and Hunter (2002, 2003) found that after 24 and 36 months, respectively, of litter additions by individual oak phenotypes into experimental microcosms, there was no effect of litter phenotype on either soil microbial biomass or microbial community composition.

A study that examined the effects of both above- and belowground inputs (*i.e.*, leaf litter and roots) across replicates of two clones found significant clonal effects on total PLFA and on ectomycorrhizal colonization (averaged across the experimental treatments; Kasurinen et al. 2005).

The latter studies support the idea that clonal variation in belowground properties may impact soil microbial characteristics, which is supported by our data.

Similarly, in agricultural systems, variation in the soil microbial community size or composition has been correlated with plant cultivar (Smith and Goodman 1999, Diab el Arab et al. 2001, Mazzola and Gu 2002). Using clonal replicates of 20 plant genotypes (across 4 cross types) we found that not only is clonal variation important to some microbial characteristics but that sometimes there is predicable variation in the community signature.

Heritability of Microbial Dynamics

Heritability of community composition was recently empirically demonstrated and mathematically defined as H^2C in Shuster et al. (2006). Based on theory outlined by Shuster et al. (2006), the tendency of replicated individuals of the same genotype to support the same biomass or community of microorganisms in soils directly beneath that tree (*i.e.*, community heritability) suggests that there are correlated fitness effects arising from interspecific indirect genetic effects (IIGEs) between the associated host plant and dependent interacting species in a community.

However, just like traditional studies of phenotypic heritability, the amount of variance explained by plant genetic factors may be largely context dependent.

Inter-specific indirect genetic effects provide the theoretical basis for individual plant genotypes having heritable community phenotypes such as what has been observed with arthropod communities (Bailey et al. 2006, Shuster et al. 2006) and now some microbial characteristics. For example, Shuster et al. (2006) found significant heritable variation in arthropod communities within P. fremontii, P. angustifolia and their backcross hybrids. Similarly, here we found significant heritable variation in microbial community composition, but only within P. angustifolia. These data suggest that there may be feedbacks at the genotype level to associated soil microorganisms (and potentially back to the plants), even though our test of microbial community composition (PLFA) may have been too coarse to detect differences that may actually exist.

IIGEs account for the interactions of the whole community (*i.e.*, the plant and each microbial community member) in which fitness of individual members is affected by their interactions within the community. While one may be able to identify all of the interacting phenotypes and their fitness consequences within a simple two or three species community, identifying all of the interacting phenotypes in a complex community is just not tractable, particularly in a microbial system.

Therefore all that can be said at present is that IIGE represent the likely mechanism for the heritable variation in P. angustifolia genotypes microbial community composition patterns (but see the following section).

Above- and Belowground Mechanisms

Our demonstration of broad-sense heritability, or clonal repeatability, in microbial biomass and community composition demonstrates that soil microorganisms can consistently respond (within a genotype) to the environment related to intraspecific genetic variation. Whether the plant genetic factors affecting microbial dynamics are related to above- or belowground processes is currently unknown. Previous work in this Populus system have shown that tree cross types (and genotypes) differ in many traits that could impact associated soil microorganisms. For example, plant secondary chemistry (condensed tannins, phenolics) and nutrients, differ between the foliage of different cross types. P. fremontii and F1 hybrids typically demonstrate low concentrations of condensed tannin while the backcross hybrids and P. angustifolia genotypes generally have high concentrations of condensed tannin in their leaf, twig and bark tissues.

Furthermore the concentration of foliar condensed tannin demonstrates broad-sense heritability within P. angustifolia (Bailey et al. 2006). Similarly, fine root production, soil CO_2 efflux and total belowground carbon allocation (BGCA; Fischer et al. 2007) and aboveground net primary productivity (AGNPP; Lojewski et al. in review) differ across the hybridizing complex and demonstrate broad-sense heritability in aboveground net primary productivity (Lowjewski et al., in review).

Trees producing high condensed tannin in their tissues have higher fine root biomass but lower overall TBCA and AGNPP than trees with low condensed tannins in their tissues. It is likely a combination of these genetic-based above- and belowground trait differences that result in specific microbial phenotypes across replicate clones, especially with respect to microbial biomass.

As Madritch and Hunter (2002, 2003) showed no effect of a 3 year litter phenotype manipulation, and a similar short-term litter manipulation experiment in the Populus system also show no effect of litter cross type (Fisher unpublished data), variation in belowground traits could be responsible for the patterns in microbial dynamics that we documented.

MANAGEMENT EFFECTS ON SOIL BIOTA

What happens to the organisms that live in the soil when land is managed for agriculture, forestry and mining? The production of food, fibre, wood products and mineral ores sometimes involves profound disturbances to the habitat of soil organisms. We clear the land of trees for grazing or cropping, we selectively log trees in forests or clear fell forests, we physically disturb the soil through ploughing or digging, we replace native plants with exotic ones, large herbivores graze the land, we use agrochemicals for suppressing weeds, invertebrate and microbial pests of plants and the parasites of livestock. How do the activities listed below affect soil biota? This photo shows cotton strips which have been buried for 17 and 44 days and a fresh, unburied strip. The microbes have been using the cellulose fibres in the cloth as a source of energy and, by 44 days in the soil, have started to rot the strip completely away in spots. You can also see how the colonies of bacteria and fungi colonising the strips have progressively stained the strips over time.

CHARACTERISTICS OF SOILS TAXOMONY

According to the United States Department of Agriculture's *Soil Taxomony*, the soils of the world fall into 12 major categories, called soil orders. Soils within one soil order share similar characteristics. With 11 of the 12 soil orders, has more soil orders than any other state in the United States. Within the 1,159 mi^2 land area alone, there are 7 different soil orders. In comparison, the entire state of Maine, which encompasses 35,387 mi^2, has only four soil oders. Knowledge of soil behaviour is very important in nutrient management because this knowledge enables you to predict how your management strategies will affect your soil.

Volcanic ash soils (Andisol)

Typical Characteristics:

- *Parent Material:* Volcanic soils, also known as Andisols, are formed from volcanic ash and cinder deposits. While most of the world's volcanic soils are not highly weathered.
- *Mineralogy:* Volcanic soils largely consist of non-crystalline (amorphous) minerals, such as allophone and imogolite.
  - These minerals form strong bonds with organic matter. As a result, organic matter generally accumulates in the surface horizon.
  - In addition to organic matter, volcanic soils may also contain high amounts of volcanic glass material with the possibility of amorphous iron and aluminum minerals.
- *Physical Traits:* Andisols are usually light and fluffy and are easily tilled. Like a sponge, these soils also hold a lot of water.
- *Fertility:* When not highly weathered, volcanic soils are typically very fertile soils. However, volcanic soils form strong complexes with phosphorus. When poorly managed, phosphorus can be limiting. Additionally, the amorphous minerals that dominate volcanic soil can generate an anion exchange capacity (AEC) when under acidic conditions and depleted in organic matter. Fertility problems may be corrected with additions of organic matter, lime and/or fertilizer amendments.

Volcanic Soils:

Glassy, dry volcanic ash soil (Vitritorrand):

- Alae Series
- Oanapuka Series

Hot, dry volcanic ash soil (Haplotorrand):

- Makena series

Moist/dry volcanic ash soil (Haplustand):

- Io Series
- Kaipoipoi Series
- Kula Series
- Laumaia Series
- Oli Series
- Olinda Series
- Pane Series

- Puu Pa Series
- Ulupalakua Series

Volcanic ash soil with very high water retention (Hydrudand):

- Hana Series
- Kailua Series
- Honomanu Series

Unique Characteristics of Volcanic Ash Soils:

Although there are several different types of volcanic soils, we will discuss two soils that are agriculturally significant.

- Moderately weathered volcanic ash soil
- Highly weathered volcanic ash soil.

The Kula Series is a prime example of a moderately weathered volcanic soil that is fertile.:

- *Location*: The Kula series can be found in the uplands near Waiakoa. In total, this series occupies approximately 8,800 acres of land Island. The approximate elevation is between 2,000 and 3,500 feet.
- *Texture*: Loam
- *Rainfall and Temperature*: The mean annual rainfall for the Kula Series is between 25 and 40 inches, and the mean annual temperature is 66 degrees F.
- *Slope*: Gentle to steep slopes
- *Parent material*: The parent material of the Kula Series is volcanic ash that was deposited above moderately weathered andesite and basalt.
- *Depth*: The typical soil layer extends 8 inches, while the depth of the subsoil is 46 inches.
- *Physical Traits*: Kula soil has good drainage, with moderately rapid permeability and medium run-off.
- *Colour of surface horizon*: Dark reddish-brown
- *Fertility*: While the Kula Series tends to have a high CEC and nutrient availability, irrigation may be required.
- *Acidity*: Slightly acid to neutral
- *Natural vegetation*: Naturally, the Kula Series supports grasses and native tree species.
- *Crops*: In cultivation, the Kula Series is used for the production of vegetable crops, flower and pastureland.
- *Related soil series*: Kaipoioi, Pane, and Kamaole series.

Depth (cm)	% C	% N	Ca (cmol/kg)	Mg (cmol/kg)	K (cmol/kg)	Sum cations (cmol/kg)
0-9	17.87	1.547	33.8	13.8	4.8	106.7
9-15	14.28	1.328	34.5	9.5	4.1	96.4
15-62	13.92	0.676	3.7	0.8	1.2	146.5
62-98	13.13		5.9	1.1	3.5	149.1
98-130	11.55		12.7	0.7	2.4	132.0

Depth (cm)	Sum Bases (cmol/kg)	Extr. Al (cmol/kg)	ECEC (cmol/kg)	Base Sat (%)	Al Sat (%)	pH (water)	pH (KCl)
0-9	52.4			49		6.3	5.5
9-15	48.1			50		6.2	5.5
15-62	5.7	3.4	9.1	4	37	5.6	4.7
62-98	10.5	5.2	15.7	7	33	5.5	4.6
98-130	15.8	3.1	18.9	12	16	5.5	4.7

High

The Hana Series is a characteristic highly weathered, infertile volcanic soil:

- *Location*: The elevation at which you will find this soil is between sea level and in the uplands of 1,200 ft.
- *Texture*: Silty clay loam
- *Rainfall and Temperature*: The average annual rainfall is between 80 and 150 inches, and the average temperature is 73 degrees F.
- *Slope*: Gently to moderately steep
- *Parent Material*: The Hana Series was derived from volcanic ash that was deposited over cinders and 'a'a lava flows. The depth of the soil profile may extend from 27 to 40 inches.
- *Depth*: While the surface soil is typically 12 inches, the subsoil extends 22 inches.
- *Physical Traits*: The surface may be very rocky depending upon the location. The soil has good drainage, with moderately rapid permeability and medium run-off.
- *Colour of surface horizon*: Dark brown/dark grayish brown
- *Fertility*: The Hana Series is typically deficient in nutrients such as phosphorus.
- *Acidity*: Strongly to medium acidity
- *Natural vegetation*: Grasses and tree species.
- *Agriculture*: Pasture.
- *Related Soil Series*: Honomanu Series

Highly Weathered Volcanic Soil vs. Moderately Weathered Volcanic Soil

Honomanu Series				Kula Series			
Depth (cm)	Ca	pH water	ECEC	Depth (cm)	Ca	pH water	ECEC
0-5	1.9	4.1		0-9	33.8	6.3	
5-14	0.3	3.8	2.3	9-15	34.5	6.2	
14-32	0.1	4.2	0.7	15-62	3.7	5.6	9.1
32-53	0.1	4.8	0.6	62-98	5.9	5.5	15.7
53-68	0.1	5.3		98-130	12.7	5.5	18.9

The major differences between a highly weathered soil, such as the Honomanu Series (Hana's sister series) and a moderately weathered soil, such as the Kula Series is highlighted in the following table. The Honomanu Series is acidic and depleted in plant nutrients such as calcium, whereas the Kula Series is slightly acidic and rich in calcium.

ARID SOILS (ARIDISOL)

Typical Characteristics

- *Distribution*: Arid soils are one of the most prevalent soil orders of the world.
- *Climate:* Arid soils are most characterised by their water deficiencies. Most arid soils contain sufficient amounts of water to support plant growth for no more than 90 consecutive days.
- *Mineralogy*: Arid soils typically contain high levels of calcium carbonates, gypsum, as well as sodium.
  - During soil formation, soluble salts at translocated to a subsurface horizon. However, since there is insufficient rainfall to leach salts from the soil profile, soluble salts accumulate.
  - Aridisols often have a surface horizon that is pale in colour and low in organic matter.
- Fertility: Due to limited moisture and accumulated soluble salts, these soils are generally not suited for major crop production. However, if properly managed and irrigated, these soil may become productive. There are notable soils that are classified as arid soils, but are notable for their unique fertility.

Arid Soils

Water saturated, salty arid soils (Aquisalid):

- Kealia Series

Minimal typical arid soil (Haplocambid):

- Keahua Series

Unique Characteristics of Arid Soils:

Of the two types of arid soils, the agriculturally significant soil is the Keahua Series.

- *Location*: The Keahua Series is lies within the central isthmus in the uplands, at elevations ranging from 600 to 1,500 ft. It extends 13,000 acres.
- *Texture*: Silty clay loam
- *Temperature and Rainfall*: Its average annual temperature is 73 degrees F, and the average annual rainfall is between 15 to 25 inches.
- *Slope*: Gentle to moderate
- *Parent material*: Keahua Series formed from basic igneous rock with some volcanic ash.

- *Depth*: Surface layer is generally 10 inches. Subsoil extends 50 inches in depth.
- *Physical Traits*: The soil is well drained and characterised by medium run-off and moderate permeability.
- *Colour of Surface Horizon*: Dark reddish brown
- *Fertility*: If irrigated, Keahua soils contain plant nutrients and are highly suitable for crop production.
- *Acidity*: Slightly acidic
- *Natural Vegetation*: Grasses, ilima, kiawe, and lantana
- *Crops*: Sugarcane, pineapple, pasture

As demonstrated in the following data table, although the Keahua Series is high in base cations, its organic carbon levels are not high enough to meet the criteria of a Mollisol, or highly fertile grassland soils.

Depth (cm)	Ca	Mg	Na	K	Org. C	Sum Cations	Base Sat	pH water
0-31	15.5	5.4	0.9	1.2	3.98*	52.4	44	6.1
31-46	11.2	5.2	0.9	0.3	2.05	41.4	43	6.6
46-58	4.6	3.5	0.8	0.1	1.01	27.3	33	6.1
58-88	2.6	2.7	1.0	0.1	0.87	25.6	25	5.6
88-160	1.1	2.6	1.5	0.1	0.91	32.0	17	5.8

SOIL RHIZOPLANE

The rhizoplane is the root epidermis and outer cortex where soil particles, bacteria and fungal hyphae adhere. The functional definition is the remaining microorganisms and soil particles after the roots have been shaken vigorously in water. There are more microbes in the rhizoplane than in the more loosely assoicated rhizosphere.

This is determined by counting the number of colony forming units (CFUs) which are determined by spreading extracted soil microorganisms across an agar and counting the number of independent clusters of microorganisms. Microbes are most abundant where the integrety of the root is compromised. For this reason, rhizoplane microorganisms tend to be found on older rather than younger roots. Bacteria and fungi that live within the cells of the root are not considered a part of the rhizoplane, but instead called endophytes.

Physical Environment

Water Potential

The plant roots which the rhizosphere is associated with can effect the physical environment of the rhizosphere. As plants transpire water with more force during the day than during the night, they change the soil water potential immediately near their roots and so the rhizosphere undergoes fluctuations that the bulk soil avoids.

Soil Texture

Movement of organic matter away from the root as well as bacteria colonizing new locations occurs more readily in sandy soils than clayey soils. Sand has larger pores between each granule allowing microorganisms and exudates can travel. Therefore, the larger the granule size, the further the rhizosphere and microorganisms associated with it will extend into the surrounding soil.

Plant roots compact the soil on the short term as they grow, but once they die and decay, can actually leave soil more porous.

pH

Several factors can lower the pH in the rhizosphere. Respiration leads to carbon dioxide (and eventually to bicarbonate/carbonic acid) generation. In addition to respiration of the roots themselves, the rhizosphere is very rich in carbon results in other organisms from prokaryotes to fungi to small animals living and respiring in the rhizosphere more than in the bulk soil.

The availible habitat that microbes are limited in part by pH of the soil. Fungi are found in more acidic soils than alkaline, and bacteria have a very broad pH spectrum where they can survive. The influincing effects of pH in the rhizosphere is critical in supporting a biologically diverse microbial community.

Plant-Derived Compounds

Plant-derived compounds are responsible for providing the additional carbon that allows the rhizosphere to host a large variety of organisms. These compounds fall into five categories: exudates, secretions, mucilages, mucigel, and lysates.

Exudates include surplus sugars, amino acids, and aeromatics that diffuse out of cells to the intercellular space and surrounding soil. Due to their diffusive nature, exudates are limited to compounds of low molecular weights. Secretions are byproducts of metabolic activity. Because they are actively released from the cell, secretions can be of both low and high molecular weight. When an epidermal root cell dies and is broken open, lysates from within the cell become available to the surrounding microbial community.

Mucilages are cells sloughed off the root cap as the root grows. Abrasive forces of the root against soil particulate matter is responsible for the removal of cells. These cells consist of cellulose, pectin, starch, and lignin. Mucigel is a slime coating the surface of a root that increases the connectivity between plant roots and the surrounding soil. It is more common on the main body of the root and root hairs than the tip. During dry spells, mucigels are responsible for allowing plants to continue to uptake water and nutrients.

Microbial Communities

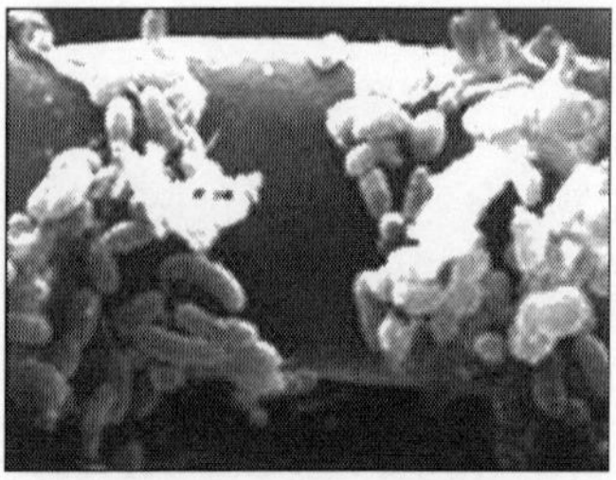

Fig. Rhizosphere Bacteria

Bacteria

Bacteria are the most numerous organisms in the soil, averaging between 10^6 to 10^9 organisms per gram of rhizosphere soil. Due to their small mass, they only account for a small amount of the biomass of soil. Nonsporulating rods, pseudomonads, and acetinomycetes are the most common bacteria in the soil.

Fungi

Both pathogenic and symbiotic fungi associate with the rhizosphere. They average between 10^5 and 10^6 organisms per gram of rhizosphere soil. Zygomycetes and hyphomycetes establish the most readily in the rhizosphere because they metabolize simple sugars.

In times of flooding, when too much moisture accumulates around the base of the tree, a bacterial pathogen Armillaria may infect and cause root rot.

Biotic Interactions in the Rhizosphere

Up to 15% of the root surface area is covered with rhizosphere-specific microorganisms - providing many sites for biological interactions. A range of interactions are present in the rhizosphere: from beneficial symbiotic relationships to detrimental pathogenic interactions.

General Impacts on Plants of Rhizosphere Microorganisms

Microorganisms in the rhizosphere complete both chemical and physical modifications to the soil profile in and around the rhizosphere that affect plants. They can be beneficial to the plant (by pathogen suppression) or detrimental (by competition for nutrients).

Chemical changes occur as a result of humification of organic matter. The resultant mineralization of various organic compounds (phosphorous, sulfur, and nitrogen, for example) provides plants with forms of nutrition that are readily available for uptake. The turnover of microbial populations also results in the release of nutrients.

The fixation of atmospheric dinitrogen by both asymbiotic and symbiotic bacteria results in increases to the available nitrogen pool that can be accessed by plants in and near the rhizosphere. Symbiotic mycorrhizae cause an increase in the effective rooting area of plants, thereby providing added nutrient mining capabilities to the plant. Rhizosphere microbes can also release plant growth regulators.

Physical changes occur primarily through the production of extracellular polymeric substances such as polysaccharides and glomalin, which improve soil aggregation and soil texture. The presence of mucigel in the rhizoplane is crucial to the water relations of plants, providing a bridge that prevents dessication by maintaining the water column during water stress events.

General Impacts on Rhizosphere Microorganisms of Plants

In addition to soil properties and the types of microorganisms present in a soil, plants play an important role in the community diversity of rhizosphere microorganisms. Plant roots cause chemical and physical changes to the soil they inhabit and these changes will affect the microbial diversity in and around the rhizosphere.

Root exudates will select for/against certain populations of microorganisms. Many plants exhibit genetic resistance/tolerance to rhizosphere microorganisms; the variety of plant will determine, in part, the community makeup of the microorganisms in the rhizospere. The ability of a plant to form symbiotic relationships with soil microbes will also determine rhizosphere microbial populations. The age and health of the plants present will also play a role in the microbial community dynamics of the rhizosphere. Plant roots increase the tilth of a soil and subsequently affect the physical properties of the soil.

Plants can also compete with rhizosphere microorganisms for resources like water and nutrients.

Symbiotic or Mutualistic Relationships

In a relationship, the plant usually provides a source of carbon and the bacteria or fungus fills some other more specialized function. In many cases the association is not absolutely necessary for the survival of both members, but provides significant benefit.

The recently sequenced genome of *Laccaria Bicolor* (an ectomycorrhizal fungus) shows a excess of enzymes used in ammonia uptake, and a lack of enzymes that would be needed to degrade plant wall material. Thus, it is assumed the fungus is providing nitrogen (and perhaps other nutrients), and a guarantee not to attack to the tree in return for glucose. As described below, fungi are also often an important source of phosphorous. The microbial partner can also help the plant to survive in an otherwise inhospitable environment. For example, specific endophytes have been shown to confer heat resistance to grasses that grow in a hydrothermal area, or salt tolerance to costal grasses.

Mycorrhizal Fungi

Over 80 percent of all land plants have a mutualistic relationship with one or more mycorrhizal fungi. Mycorrhizal fungi extend the effective root length by 100 fold or more. It would therefore seem appropiate that mycorrhiza means fungus-root in Greek.

Not only do they increase the surface area over which nutrients and water can be taken up by the plant, but fungal hyphae can reach into smaller pores than roots can. The most valuable asset mycorrhizae provide is accessing immobile nutrients, such as phospherous. In return, the plant passes a significant amount of its carbon reserves to the fungi.

PLANT GROWTH-PROMOTING RHIZOBACTERIA (PGPR)

The concept of PGPR is now well established and so some consideration of the relationship of PGPRs to biocontrol is worthwhile. PGPR increase plant growth indirectly either by the suppression of well-known diseases caused by major pathogens or by reducing the deleterious effects of minor pathogens (micro-organisms which reduce plant growth but without obvious symptoms). Most of the bacteria discussed so far in this review fall into this category of PGPR.

Alternatively, PGPR may increase plant growth in other ways, for example, by associative N2 fixation (Hong et al., 1991), solubilizing nutrients such as P (Whitelaw, 2000), promoting mycorrhizal function (Garbaye, 1994), regulating ethylene production in roots (Glick, 1995), releasing phytohormones, and decreasing heavy metal toxicity (Burd et al., 1998).

It has been suggested that the two groups should be reclassified into biocontrol plant growth-promoting bacteria (biocontrol PGPB) and PGPB (Bashan and Holguin, 1998). To date this proposal does not seem to have been widely accepted, but it does highlight the need to consider the full ecological interactions taking place following application of bacteria to seeds and roots that lead to plant growth promotion.

It is also important to remember that deleterious rhizobacteria that inhibit plant growth are also known (Nehl et al., 1996) which can influence such interactions.

Irrespective of mode of action, a key feature of all PGPR is that they all colonize roots to some extent. In some cases this may involve specific attachment through, for example, pili, as with the attachment ofPseudomonas fluorescens 2-79 to the surface of wheat roots (Vesper, 1987). However, such specific attachment does not seem to be an absolute requirement for colonization (de Weger et al., 1995). Colonization may involve simply root surface development but, endophytic colonization of the root is also known, and the degree of endophytic colonization depends on bacterial strain and plant type. Endophytic growth in roots has been recorded with the PGPR Bacillus polymyxa (Prazmowski) Macé Pw-ZR and Pseudomonas fluorescens Sm3-RN on spruce (Picea glauca×P. engelmannii) (Shishido et al., 1999), with the biocontrol strains of Bacillus sp. L324-92R12 and P. fluorescens 2-79RN10 on wheat (Kim et al., 1997) and with several that induce resistance such as Bacillus pumilus Meyer and Gottheil SE34 and P. fluorescens 63-28 on pea (Pisum sativum L.), P. fluorescens CHA0 on tobacco (Troxler et al., 1997) and P. fluorescensWCS417r on tomato (Duijff et al., 1997). Large scale differences in spread within the plant may occur.

Some, such as B. polymyxa Pw-2B, P. fluorescens SM3-RN, Bacillus sp. L324-92R12, and P. fluorescens 2-79RN10spread from roots to aerial plant parts whereas others may not (Kim et al., 1997). Small scale differences are also known. For example, both B. pumilus SE34 and P. fluorescens 63-28 grow on the root surface and intercellarly in pea roots whereas surface growth, inter- and intra-cellular growth occurred with P. fluorescens WCS417r in tomato and P. fluorescens CHA0 in tobacco. These endophytic bacteria may be in a particularly advantageous ecological position in that they may be able to grow and compete on the root surface, but also may be capable of developing within the root, relatively protected from the competitive and high-stress environment of the soil.

Indeed, many seeds, roots and tubers are normally colonized by endophytic bacteria. Any plant resistance encountered must be minimal, although, in many cases, sufficient to allow ISR to develop. The localized signalling between plant and bacteria within the root environment deserves further study. Certainly, use of mutants and promoter probe techniques are beginning to identify genes in bacteria that are important in colonization and these are often related to nutrient uptake. Such nutrient uptake genes may also play a role in biocontrol by aiding the uptake and metabolism of nutrients that stimulate germination of pathogen propagules (Maloney et al., 1994).

The ability to colonize seeds is also an important feature for many bacterial biocontrol agents. Pseudomonas chlororaphis (Guignard and Sauvngeau) Bergey et al. MA342 is applied to cereal seeds to control many seed and soil-borne pathogens and has been found to colonize specific areas of the seed coat (Tombolini et al., 1999). After inoculation, the bacteria were found under the seed glume (or husk), but after planting they were found to colonize the glume cells epiphytically. Bacterial aggregates were also found in the grooves formed by the base of the coleoptile and the scutellum, and near the embryo but never within it.

In this case, the biocontrol bacteria co-located with the seed-borne pathogen Drechslera teres (Sacc.) Shoemaker providing biocontrol through the production of fungitoxic compounds. The spermosphere competence of this bacterium allowed biocontrol to take place. Microcolony or microaggregate production by bacteria has also been found on the grooves or cracks on the outer seed coat of sugar beet and cotton (Gossypium hirsutum L.) perhaps reflecting areas of increased nutrient availability or environmental protection.

FUNGAL–PROTOZOAN INTERACTIONS

The soil-borne protozoan Plasmodiophora brassicae Woronin is an ecologically obligate biotroph of brassicas causing clubroot disease which is characterized by proliferation of galls on infected roots. From a large-scale screening exercise, two isolates of the root-colonizing fungusHeteroconium chaetospira (Grove) Ellis were found to suppress clubroot on chinese cabbage (Brassica campestris L.) in non-sterile soil (Narisawa et al., 1998). Hyphal growth occurred in the inner parts of the cortical tissues and into the root tips without causing any external symptoms on the plant and there was no sign of infection by P. brassicae.

Further studies demonstrated that H. chaetospira infected epidermal cells from appressoria via infection pegs and, subsequently, intracellular hyphal growth occurred (Narisawa et al., 2000). However, the actual mechanism of the disease control observed in the field was unclear. Heteroconium chaetospira appears to form a mutualistic symbiosis with B. campestris in terms of disease control which is of interest as the Brassicacae family is largely non-mycorrhizal. In addition, H. chaetospira was found to colonize the roots of plants from eight families and may have a wide host range (Narisawa et al., 2000). Its ability to control diseases in these other plant families and the mechanisms involved deserves further study.

FUNGAL–BACTERIAL PATHOGEN INTERACTIONS

In the last few years there have been no clear examples of fungi used to control bacterial plant pathogens in the rhizosphere or spermosphere. The reasons for this are unclear but could perhaps indicate an area that deserves further research in the future.

FUNGAL–FUNGAL PATHOGEN INTERACTIONS

Interactions between biocontrol fungi and fungal plant pathogens continue to be the focus of a large number of researchers, on a par with work on bacterial–fungal plant pathogen interactions described earlier. However, there is an extra dimension in the quality of the interactions between fungi as biocontrol fungi have much greater potential than bacteria to grow and spread through soil and in the rhizosphere through possession of hyphal growth. Some recent examples of fungal–fungal interaction concerning biocontrol in the rhizosphere and spermosphere.

There are a variety of fungal species and isolates that have been examined as biocontrol agents but Trichoderma species clearly dominate, perhaps reflecting their ease of growth and wide host range. There has been an upsurge in interest in non-pathogenic Pythium species, particularly P. oligandrum Drechsler where additional modes of action have been determined recently, and a continued interest in well-established saprotrophic antagonists such as non-pathogenic Fusarium species, non-pathogenic binucleate Rhizoctoniaisolates and Phialophora species, as well as mutualistic symbionts including mycorrhizal fungi such as Glomus intraradices Schenk and Smith. At least one novel biocontrol agent, Cladorrhinum foecundissimumSaccardo and Mardial, has been described.

Numerous others are listed elsewhere (Whipps, 1997a). The most common pathogen targets arePythium species, Fusarium species and Rhizoctonia solani reflecting their world-wide importance and perhaps their relative ease of control under protected cropping systems, although numerous other pathogens have been examined.

Significantly, relatively few of the examples involve studies in non-sterile soil or field conditions, with most carried out in soil-less conditions reflecting the need to keep the complexity of the system to a minimum in order to achieve reproducible control. Some specific examples of the modes of action found to occur in the rhizosphere and spermosphere during interactions between fungi and fungal plant pathogens.

Competition

There have been relatively few studies on competition for nutrients, space or infection sites between fungi in the rhizosphere and spermosphere recently. Competition for carbon, nitrogen and iron has been shown to be a mechanism associated with biocontrol or suppression of Fusarium wilt in several systems by non-pathogenic Fusarium and Trichoderma species and competition for thiamine as a significant process in the control ofGaeumannomyces graminis var. tritici by a sterile red fungus in the rhizosphere of wheat (Shankar et al., 1994). Many studies have shown a relationship between increased colonization of the rhizosphere by a non-pathogen, associated subsequently, with disease suppression.

This is well established for non-pathogenic strains of Fusarium oxysporumcontrolling pathogenic F. oxysporum on a variety of crop plants, hypovirulent or non-pathogenic binucleate strains of Rhizoctonia species to control pathogenic isolates of R. solani (Herr, 1995) and several fungi includingPhialophora species, Gaeumannomyces graminis var. graminis and Idriella bolleyi (Sprague) von Arx as well as several non-sporulating fungi, to control G. graminis var. tritici.

As just one example, I. bolleyi exploits the naturally senescing cortical cells of cereal roots during the early stages of the crop and outcompetes G.

graminis var. tritici for infection sites and nutrients. Rapid production of spores, which are then carried down the root by water, continue the root colonization process and this is suggested to be a key feature in the establishment of the biocontrol agent on the root.

Mycorrhizal fungi are also strong candidates for providing biocontrol through competition for space by virtue of their ecologically obligate association with roots. Ectomycorrhizal fungi because of their physical sheathing morphology may well occupy normal pathogen infection sites. Strangely, little work has been carried out to demonstrate this mechanism since it was first suggested (Marx, 1972) with most biocontrol interest focused on antibiotic production and induced resistance (Perrin, 1990;Duschesne, 1994).

Similarly, arbuscular mycorrhizas also have potential to occupy space and infection sites on roots, but evidence suggests that biocontrol provided by arbuscular mycorrhizas relates more to induced resistance, improved plant growth and changes in root morphology rather than competition per se.

Antibiosis

Although production of antibiotics by fungi involved in biocontrol is a well-documented phenomenon, there is little recent work clearly demonstrating production of antibiotics by fungi in the rhizosphere and spermosphere. Unlike the situation with biocontrol bacteria, there appear to be no detailed studies in biocontrol fungi of genes coding for antibiotic synthesis.

Mutants with raised or decreased production of antibiotics are either natural spontaneous ones or generated by UV or chemical mutagenesis, with inherent problems of pleiotropic gene effects, rather than targeted gene disruption. Consequently, clear identification and understanding of the role of antibiotics in disease control lags far behind that in bacteria and needs to be addressed.

Antibiotic production by fungi exhibiting biocontrol activity has most commonly been reported for isolates of Trichoderma/Gliocladium (Howell, 1998) and Talaromyces flavus (Klöcker) Stolk and Samson although in the last few years antibiotics have been at least partially characterized in Chaetomium globosum (Kunze) (Di Pietro et al., 1992). Minimedusa polyspora (J. W. Hotson) Weresub and Le Clair (Beale and Pitt, 1995) and Verticillium biguttatum Gams (Morris et al., 1995). Of particular interest are those studies where antibiotic production has a definite link to biocontrol. For example, Trichoderma (Gliocladium)virens (J. Miller, Giddens and Foster) von Arx comprises P and Q group strains, based on their antibiotic profiles (Howell, 1999).

Strains of P group produce the antibiotic gliovirin which is active against Pythium ultimum but not against Rhizoctonia solani AG-4. Strains of the Q group produce the antibiotic gliotoxin which is very active against R. solani but less so against P. ultimum. In seedling bioassay tests, strains of the P group are more effective biocontrol agents of damping-off on cotton caused by Pythium, while those from the Q group are more effective as biocontrol agents of damping-off incited by R. solani (Howell, 1991;Howell et al., 1993). Thus there is strong circumstantial evidence for a role for antibiotics in biocontrol in this experimental system.

This has been confirmed in a zinnia–Pythium system where T. virens G-20 incorporated into soil and potting mix resulted in disease suppression clearly associated with maximum accumulation of gliotoxin in the medium. Gliotoxin minus mutants displayed only 54 per cent of the Pythium disease suppressive activity in zinnia compared with the wild-type (Wilhite et al., 1994). Gliotoxin production by Trichoderma is also thought to be responsible for cytoplasmic leakage from R. solani observed directly on membranes in potting mix (Harris and Lumsden, 1997).

Production of hydrogen peroxide in the rhizosphere, catalysed by glucose oxidase from Talaromyces flavus is thought to be responsible for the biocontrol of Verticillium wilt caused by Verticillium dahliae Kleb. on eggplant (Solanum tuberosum L.) (Stosz et al., 1996). Purified glucose oxidase significantly reduced the growth rate of V. dahliae in the presence, but not the absence, of eggplant roots, suggesting that a supply of glucose from the roots was of major importance (Fravel and Roberts, 1991).

Further, a single-spored variant, Tf-l-np, which produced 2 per cent of the level of glucose oxidase activity of the wild-type did not controlVerticillium wilt on eggplant in non-sterile field soil in a glasshouse experiment, whilst the wild-type provided significant control (Fravel and Roberts, 1991). Glucose oxidase also suppressed growth of V. dahliae in vitro and killed microsclerotia of V. dahliae in vitro and in soil.

Induced Resistance

As with bacteria described earlier, the ability of fungi to induce resistance in plants and provide biocontrol has gradually been receiving more attention in the last few years. A considerable number of fungi previously described to provide biocontrol by mechanisms such as competition, antibiosis, mycoparasitism or direct growth promotion are now thought to provide control, at least in part, by this mechanism. These include saprotrophs such as non-pathogenic Fusarium isolates, Trichoderma species (Yedidia et al., 1999), Pythium oligandrum, non-pathogenic binucleate Rhizoctonia isolates, and Penicillium oxalicum Currie and Thom (de Cal et al., 1997) as well as mutualistic biotrophs such as mycorrhizal fungi.

However, not all these studies used the strict criterion of spatial separation between application of the biocontrol fungus and the challenging pathogen to define induced resistance. Indeed, some simply measured changes in enzymes, PR-proteins or cell wall characteristics found to be induced in plants through SAR (described earlier) without involvement of a pathogen at all. Certainly with some mycorrhizal fungi it has been questioned whether the biochemical responses similar to induced resistance found following infection are of sufficient magnitude or quality, or too transient, to provide disease control.

Indeed, during some mycorrhizal syntheses there is little or no induced resistance response detected (Mohr et al., 1998). However, spatial or temporal separation experiments have indicated that increased levels of chitinases, β-1,3 glucanases, β-1,4 glucosidase, PR-1 protein, and peroxidase as well as cell wall appositions and phenolics may be associated with induced resistance due to fungi.

Nevertheless, more work is needed to identify the biochemical changes taking place in a larger number of fungal–plant combinations as not all these biochemical markers were found to be important in each system examined. Further, there appear to be differences in the quality of the induced resistance found between bacteria and on fungi on the same plant. For example, the suppression of Fusarium wilt (F. oxysporum f. sp. lycopersici) (Sacc.) Snyder and Hansen on tomato by Pseudomonas fluorescens WCS417r did not involve production of PR-1 protein and chitinases whereas that induced by F. oxysporum Fo47 did (Duijff et al., 1998). Again more work in this area is required to determine the extent of differences in induced resistance produced by bacteria and fungi.

The elicitors responsible for inducing resistance are not known in detail.Trichoderma species produce a 22?kDa xylanase that, when injected in plant tissues, will induce plant defence responses including K+, H+ and Ca2+ channelling, PR protein synthesis, ethylene biosynthesis, and glycosylation and fatty acylation of phytosterols (Bailey and Lumsden, 1998). However, whether such a system is active in roots exposed toTrichoderma is not known.

Pectic oligogalacturonides released after hydrolysis by a non-pathogenic binucleate Rhizoctonia isolate may act as elicitors of defence responses in bean (Phaseolus vulgaris L.) (Jabaji-Hareet al., 1999).

Dose–response experiments involving non-pathogenic Fusarium species to control F. oxysporum on tomato have indicated that induced resistance is not an all or nothing response (Larkin and Fravel, 1999). By varying the level of inoculum of the inducing strain and the pathogenic isolate in soil, it was shown that some non-pathogenic isolates such as Fusarium CS-20 controlled Fusarium wilt effectively with antagonist levels of only 100 chlamydospores g-1 of soil (cgs) with pathogen densities of up to 105?cgs.

In contrast, isolate Fo47 was effective only at antagonist densities of 104–105 cgs, regardless of pathogen density. Subsequent mathematical modelling provided evidence that CS-20 control was largely through induced resistance whereas Fo47 was active primarily through competition for nutrients (Larkin and Fravel, 1999). Similar dose–response effects were found with non-pathogenic isolate of F. oxysporum f. sp.ciceris (Padw.) Matuo and Sato and non-pathogenic isolates of F. oxysporumto control wilt of chickpea (Cicer arietinum L.) caused by pathogenic F. oxysporum f. sp. ciceris (Hervás et al., 1995). However, in addition, the plant genotype also seemed to influence the degree of resistance induced.

Mycoparasitism

There is a huge literature on the ability of fungi to parasitize spores, sclerotia, hyphae, and other fungal structures and many of these observations are linked with biocontrol. However, most of the microscopical observations concerning mycoparasitism have come from in vitro studies or sterile systems and examples clearly demonstrating mycoparasitism in the rhizosphere or spermosphere are rare (Lo et al., 1998).

However, indirect population dynamic studies showed that mycelium of Rhizoctonia solani in the rhizosphere of potato was a prerequisite for development of the mycoparasite Verticillium biguttatum and rhizosphere competence was strongly related to biocontrol in mycoparasite isolates of Trichodermaspecies.

The process involved in mycoparasitism may consist of sensing the host, followed by directed growth, contact, recognition, attachment, penetration, and exit. Although not all these features occur in every fungal–fungal interaction, the key factor is nutrient transfer from host to mycoparasite. Directed growth of hyphae of Trichoderma to hyphae ofRhizoctonia solani prior to penetration has often been observed (Chet et al., 1981) and the presence of host sclerotia have been shown to stimulate germination of conidia of Coniothyrium minitans Campbell (Whipps et al., 1991) and Sporidesmium sclerotivorum Uecker, Adams and Ayers.

However, the factors involved in controlling directed growth in these systems have not been fully characterized. Similarly, the factors controlling recognition and binding between fungal host and parasite are not yet clear. This process may involve hydrophobic interactions or interactions between complementary molecules present on the surface of both the host and the mycoparasite such as between lectins and carbohydrates. With Trichoderma, there is good evidence of lectin production by both parasite and host Corticum(Sclerotium) rolfsii Curzi and involvement of lectins in the differentiation of mycoparasitism-related structures.

Recently, both hydrophobic characteristics and surface carbohydrate moieties have been investigated in the mycoparasite C. minitans as a prerequisite to examining the interaction with its hostSclerotinia sclerotiorum (Lib.) de Bary (Smith et al., 1998, 1999). Little is known of the signalling pathways following recognition of the host. However, preliminary evidence in Trichoderma harzianum indicates that the signal is transduced by heterotrimeric G proteins and mediated by cAMP (Omero et al., 1999). As penetration or cell wall degradation are frequently observed during mycoparasitism, great emphasis has been placed on characterizing and cloning the extracellular enzymes such as β-1,3 glucanases, chitinases, cellulases, and proteases produced by fungal biocontrol strains.

By manipulating their activity through construction of 'overproducing' mutants, enzyme-negative mutants or even transgenic plants expressing the enzyme, a role for their production in biocontrol has been implied. Several fungi have been examined in this way including Talaromyces flavus (Madi et al., 1997), but this type of work has essentially focused on Trichoderma species. For example, a series of transformants of T. longibrachiatum Rifai were constructed with extra copies of egl1 gene encoding the production of β-1,4 endoglucanase (Migheli et al., 1998).

When applied to cucumber seeds sown in Pythium ultimum-infested soil, the transformants with inducible or constitutive expression of egl1 were generally more suppressive than the wild type strain. In this case, it was suggested that P. ultimum was controlled through the action of β-1,4 glucanase degrading the cellulose of the cell wall of the pathogen.

Similarly, transformants of T. harzianum, overproducing proteinase encoded by prb1, provided up to a 5-fold increase in control of damping-off in cotton caused by Rhizoctonia solani(Flores et al., 1996). Interestingly, the best protection was provided by a strain which produced only an intermediate level of proteinase activity and it was suggested that very high levels of proteinase production might cause degradation of other enzymes which are important in the mycoparasitic process (Flores et al., 1996). In this regard, chitinases have received the greatest attention in mycoparasitism.

Numerous studies have been made of β-N-acetylhexosaminidase (EC 3.2.1.52) (which splits the chitin polymer into N-acetylglucosamine monomers in an exo-manner), endochitinase (EC 3.2.1.14) (which cleaves randomly at internal sites over the entire length of the chitin microfibril) and chitin 1,4-β-chitobiosidase (exochitinases or chitobiosidases) (which releases diacetylchitobiose in a stepwise fashion such that no monosaccharides or oligosaccharides are formed). For example, transformants of T. harzianum CECT 2413 that over-expressed on 33 kDa endochitinase (chit33) were more effective in inhibiting the growth of Rhizoctonia solani in vitro compared with the wild type (Limónet al., 1999).

The combination of chitinases as well as other cell wall-degrading enzymes differ between species and strains (Lorito, 1998) and chitinases are differently expressed during mycoparasitism. For example, an N-acetylhexosaminidase (CHIT 102) was the first to be induced in T. harzianum T-Y, but as early as 12 h after contact with its host Sclerotium rolfsii, its activity diminished, while that of another N-acetylhexosaminidase increased (Haran et al. 1996b). In contrast, whenRhizoctonia solani was the host, CHIT 102 was stimulated along with three endochitinases within 12 h following contact but, as the interaction proceeded, CHIT 102 activity decreased and that of the endochitinases increased (Haran et al., 1996b).

In T. (atroviride) harzianum P1, expression of endochitinase ech42 occurred before contact with the hostR. solani when a N-acetylhexosaminidase nag1 was not induced until after contact (Zeilinger et al., 1999). Degradation products of the cell wall were considered to act as inducers of these enzymes in this host–pathogen system.

Interestingly, when examining T. harzianum P1 interaction withBotrytis cinerea Pers., ech42 gene transcription was found to be triggered by physiological stress reflecting carbon source depletion rather than the presence of chitin as the inducer (Mach et al., 1999). Significantly, ech42minus mutants did not reduce biocontrol activity against Pythium ultimumor S. rolfsii and appeared to enhance activity against R. solani, suggesting that expression of this gene in T. harzianum P1 was not directly important for the biocontrol of these fungi.

In contrast, activity against B. cinerea was impaired, clearly indicating that the mechanism involved in antagonistic interactions with T. harzianum P1 differs with host, and is unlikely to depend solely on the production of a single cell wall-degrading enzyme. Indeed, evidence from studies involving combinations of purified cell wall-degrading enzymes and antibiotics from Trichoderma species support this idea.

The final evidence for a role for cell wall-degrading enzymes in biocontrol involves the expression of fungal genes in transgenic plants. For example, an endochitinase from Trichoderma harzianum has been transformed into tobacco and potato and the transgenic plants showed a high level of resistance to a broad spectrum of diseases (Lorito, 1998). Similarly, transgenic apple trees expressing an endochitinase from T. harzianumalso exhibited increased resistance to apple scab caused by Venturia inaequalis (Cooke) Winter, although plant growth was reduced (Bolar et al., 2000). The potential consequently exists to combine different enzymes in transgenic plants to obtain synergistic biocontrol and these experiments are underway.

One feature often overlooked with mycoparasitism is that it may not always be confined to control of plant pathogens. A mycoparasite may also have the potential to attack beneficial fungi such as those forming mycorrhiza. For example, T. harzianum T-203 was shown to attack mycelium of the arbuscular mycorrhizal fungus Glomus intraradices in an axenic system (Rousseau et al., 1996).

However, in a soil-based system, G. intraradices was unaffected by the presence of T. harzianum T3a. IndeedT. harzianum appeared to be suppressed through nutrient competition (Green et al., 1999). In contrast, in a different in vitro system, soluble material obtained from G. intraradices mycelium stimulated conidial germination of T. harzianum, but not that of the pathogen

F. oxysporumf. sp. chrysanthemi G. M. and J. K. Armstrong and Littrell (Filion et al., 1999). These results may reflect the different isolates of fungi and plants used, and the experimental systems applied, but they clearly demonstrate the complexity of the interactions that can occur in the rhizosphere.

CONTROL MEASURES FOR SOIL-BORNE PATHOGENS

Pythium Soft Rot was widespread and occurred in all of the major ginger growing areas in Fiji. It has resting oospores that can still be infectious even after long fallows. It is also transferred on infected planting material as our studies confirmed. R. similis, on the other hand, occurred on a few farms, and only in the Veikoba region. The nematode was sometimes not detected in soil at the time ginger was planted, but was invariably introduced in planting material. It then multiplied rapidly and was readily detected later in the season.

The highest population densities recorded from ginger fields were 170 nematodes/200 ml soil at immature harvest and 410 nematodes/200 ml soil at mature harvest. R. similis was never found in taro or cassava roots, but was occasionally found at low population densities in soil associated with roots of those crops. However, moderate to high populations were invariably found on volunteer ginger, which was always present as a weed in the following taro and cassava crops.

Soils amended with poultry manure and sawdust (PS) and cropped soils that were subject to minimal disturbance (MT) were most suppressive to Pythium Soft Rot and plant parasitic nematodes. Conversely, treatments where the soil microbial community was disturbed or diminished (*i.e.* fumigated and bare fallow soils) had significantly lower levels of suppressiveness compared to the Crop/MT/PS treatment.

These results demonstrate that the way soil is managed influences its suppressiveness to soil-borne pathogens. It should therefore be possible to reduce the impact of these pathogens by modifying the current ginger farming system. Furthermore, in the Fijian ginger farming system, ginger is generally grown every 3 years, with crops of cassava and taro planted in the years between ginger crops. Our results indicate that this is an excellent rotation for managing both Pythium Soft Rot and plant parasitic nematodes. Taro and cassava are poor hosts for both of these pathogens. Even in the Australian context, a 3-year break from ginger is still likely to be useful in managing soil-borne pathogens, as inoculum densities will decline during the break if non-host plants are used, and thus reduce disease severity when ginger is planted.

A framework was developed for the project activities through regular meetings with research and extension staff, as well as external consultants, and was updated annually. This process provided direction to the project and improved our understanding of the soil-borne pathogens affecting ginger production and provided growers with better options for control in two markedly different environments and cropping systems. In addition a survey of the technical efficiency and output losses (pathogens and soil erosion) of Fijian ginger growers was completed by Mr Waisiki Gonemaituba as part of his Masters studies at the University of Queensland. His study has important policy implications that the Fijian government can use to create a more profitable and sustainable ginger industry.

In terms of industry extension, project staff presented the results of the first, 2-year phase of research completed in Fiji and Australia at a seminar held at Koronivia Research Station on 13 February 2009. Research staff from the Ministry was present, as were interested SPC plant protection staff and ginger growers.

Towards the completion of the project a major workshop was held on 30 July 2010 at a local processor/exporters premises in Navua and was very well attended, drawing the support of the Minister of Agriculture and the Permanent Secretary of Agriculture, together with the Heads of Departments and influential local growers.

The workshop proved to be a catalyst for gaining the support of private business partners in establishing a clean seed scheme in areas declared free from R. similis. It also provided impetus for the printing of information leaflets in the local language and in launching field days and walks for growers in the major production areas.

In Australia, Field Days organised by the Australian Ginger Growers' Association (AGGA) were held in January for each year of the project and results of project activities were presented through field walks, oral presentations and hand-outs. Project updates have also been provided to growers at quarterly AGGA meetings. Also in Australia, with support from Queensland Primary Industry and Fisheries (QPIF) trade and investment officers, a consultative process was commenced to identify pathways by which all stakeholders within the ginger industry supply chain may increase their profitability.

The activities undertaken during the reporting period included a preliminary investigation of the ginger industry to better understand issues within the industry, which was followed up with a detailed investigation of the industry to identify market trends and opportunities for the industry. A report was made available to the Australian Ginger Growers' Association on 11 May 2009.

Forward momentum was maintained by the convening of a Ginger Industry Development Workshop on 27 May 2009 at Maroochy Research Station to develop an action plan that will guide future investment in the industry and foster its development. An industry R&D levy has since received the support of growers and processors and will be administered by the Rural Research and Development Corporation.

PYTHIUM SOFT ROT OF GINGER

Observations on Crop Losses and Disease Etiology

Our observations, together with the unpublished results of Fullerton and Harris (1998), show that losses due to Pythium Soft Rot occur regularly in Fiji. In 1997–98, rhizome rot destroyed ~30, 75, 30 and 100 per cent of mature ginger crops at four field sites in Fiji (Fullerton and Harris 1998), and we observed similar losses to seed ginger in 2007–08. We suggest that given the typically wet Fijian climate, rhizome rot is likely to occur on seed crops in most years and will often limit the amount of material available to plant the next ginger crop.

Since most Fijian ginger is now harvested in February for the immature market, total losses from the disease are probably not as great as they were 10 years ago. Nevertheless, immature ginger can suffer heavy losses when conditions are unusually wet for a month or two before harvest. Such a situation occurred in the 2008–09 season, as losses of 25–50 per cent were observed in immature crops in the Navua district following 789 mm of rain on 29 days during January 2009.

The fact that rhizome rot occurred in Australia during the time we were studying the disease in Fiji was fortuitous in some ways, as the summer of 2007–08 was the wettest in the Australian ginger industry for more than 30 years. During the 2 months from mid-December 2007 to mid-February 2008, temperatures were high and rain fell almost every day; conditions that occur regularly in Fiji.

Although severe losses from rhizome rot have never previously been observed in Australia, it is possible that the disease occurred at low levels before 2007–08 and had never been reported. Australian ginger growers normally assume that patches of yellowing plants with rotting rhizomes are affected by Fusarium yellow (caused by Fusarium oxysporum f. sp. zingiberi), as this disease occurs on almost every farm in the industry. Dohroo (2005) noted that wet conditions, high soil water and relatively low soil temperatures were the most important factors influencing the development of rhizome rot in India and our observations suggest that the first two are also important in Fiji and Australia.

Continuity of wetness may be as important for disease development as total rainfall, as the disease is clearly associated with free water and saturated soils. Under such conditions, zoospores disperse in soil water and spread the disease to adjacent clumps (Dohroo 2005), although our observations of greater downhill than uphill movement suggest that mass flow of water is also involved.

However, we question the contention of Dohroo (2005) that Pythium infection is favoured by low soil temperatures. Although this may be true for some Pythium species, P. myriotylum is most active at mid-summer temperatures. Both Fijian and Australian isolates have high optimum temperatures (radial growth rates of 28–34 mm/day on agar at 37°C) and high temperatures enhance the pathogenicity of P. myriotylum on other crops in Queensland. Although our pathogenicity experiments were performed with P. myriotylum, we also found other species of Pythium which were associated with the disease. The fact that severe rhizome rot was initially observed on only two Australian farms raises questions about factors that may have caused the disease to express itself on those farms.

Rainfall and soil type did not appear to be involved, as farms near the affected Eumundi farm received similar rainfall; the affected farm at Kandanga was slightly drier than other farms in the industry and the disease was not observed on adjacent farms with similar soil types. The most likely explanation is that the soils on the affected farms had deteriorated due to long-term cultivation.

The Eumundi and Kandanga farms were first planted to ginger in 1960 and 1978, respectively, and are therefore amongst the oldest in the Australian ginger industry and probably the most intensively cultivated. Variously classified as ferralic nitisols, red ferrosols or krasnozems (McKenzie et al. 2004), these soils were originally relatively fertile and drained readily.

However, after 30–50 years of cultivation, it is possible that drainage is now somewhat impeded due to compaction and declining levels of organic matter. Since Pythium is a relatively poor competitor in microbially active soils (Hoitink and Boehm 1999; Stone et al. 2004), a reduction in the soil's biological buffering capacity may also be a contributing factor.

Molecular Identification of Pythium isolates

BLAST analysis revealed that three Pythium species, P. myriotylum, P. vexans and P. graminicola, were identified each from three different ginger growing localities (Veikoba, Navua and upper Naitasiri, respectively). The two Australian ginger isolates revealed sequence alignment with P. myriotylum and P. zingiberis; the capsicum isolate was confirmed as P. myriotylum.

Table. Pythium isolates from Fiji (KRS suffix) and Queensland (BRIP suffix) showing location collected, species identification based on ITS sequence similarity and Genbank accession number where deposited. All isolates from ginger except* from capsicum.

Sample	location	Pythium sp.	GenBank
KRS11	Navua	P.vexans	
KRS13	Muainaweni	P.graminicola	
KRS14	Veikoba	P.myriotylum	FJ797574
KRS15	Waibau	P.graminicola	
KRS17	Veikoba	P.myriotylum	FJ797575
BRIP39907*	Bundaberg	P.myriotylum	FJ797576
BRIP52426	Eumundi	P.myriotylum	FJ797577
BRIP52427	Eumundi	P.zingiberis.	FJ797578

This is the first record of P. vexans and P. graminicola from ginger in Fiji and the first putative record of P. zingiberis in Australia. According to Dohroo (2005) P.graminicola is present in Sri Lanka while P.vexans in India.

The two species have been found to cause problems during rainy weather. The surveys are still relatively limited and there may be more species present in the Fijian ginger growing areas, or indeed other species present in Fiji and Australia with the capacity to cause rhizome rot in ginger. Consequently, more survey work is warranted. P. zingiberis has been recorded in Japan and Korea (Ichitani and Shinsu, 1980).

In Japan, it has been isolated from various parts of rotten ginger, especially from the basal part of terrestrial stem and rhizomes regardless of stages of disease development and locations. It has been isolated from soils where ginger is growing and from areas where ginger has been previously grown. Morphologically, P.zingerberis is very similar to P.myriotylum and at the molecular level only a single base pair difference in the ITS2 region of the rDNA separates the two species (Levesque and de Cock, 2004). P.zingiberis has never been recorded in Australia, and to validate this record further morphological analysis is required.

Impact of Soil Management Practices on Soil Health

In September 2005, various cropping, tillage and amendment treatments were established in a sandy clay loam soil (Grey Dermosol) on a ginger farm near Yandina. This site was free of ginger pathogens because it had previously grown sugarcane. A similar set of treatments was included in an experiment established in April 2006 on a light clay soil (Red Ferrosol) on a farm near Kandanga.

The continuous cropping treatment consisted of growing maize, soybean or forage sorghum during summer and oats or brassica during winter under two tillage regimes [conventional tillage (CT) or minimum tillage (MT)]. There was also a permanent pangola grass pasture, together with a bare fallow that was cultivated regularly to control weeds (at Yandina) or a CT treatment at Kandanga that was fumigated with 1,3 dichloropropene in 2007 and 2009. Prior to establishing the crop and pasture treatments, half the plots received a poultry manure/sawdust amendment (PS) of 100 t/ha at Yandina or 20 t/ha at Kandanga.

Soil was collected from each experiment in September 2009, a week before the site was planted to ginger. Half the soil was irradiated at 25 kGy and both irradiated (IR) and non-irradiated (NIR) soils were placed in 2 L planter bags and planted with a 60 g piece of ginger rhizome, as described previously for suppression studies. Leaf yellowing began within 2 weeks of inoculation and symptoms then increased in severity for the duration of the experiment. Less than 5 per cent of the 216 non-inoculated controls developed rhizome rot, indicating that background levels of Pythium were relatively low in the soils used for the experiment. Disease severity was much greater in irradiated than non-irradiated soil, indicating that biological factors influenced disease expression. Interestingly, the soil from Yandina (pH(water) = 4.7) was more suppressive to Pythium than the soil from Kandanga (pH(water) = 6.2). However, direct pH effects could not account for the difference in suppressiveness, as disease severity was high in both soils after they were irradiated. It is therefore possible that the soil microorganisms which were inhibiting Pythium (through competition, antibiosis or parasitism) were more active in acidic soils.

Table. Impact of irradiation and soil type on the severity of Pythium rhizome rot in ginger

Site	Irradiated	Non-irradiated
Kandanga	2.71 c	2.28 b
Yandina	2.56 bc	1.38 a

Values (disease severity ratings after 3 months) followed by the same letter are not significantly different (P = 0.01) Soils amended with PS and cropped soils that were subject to minimal disturbance (MT) were most suppressive to rhizome rot. Conversely, treatments where the soil microbial community was disturbed or diminished (*i.e.* fumigated and bare fallow soils) had significantly lower levels of suppressiveness compared to the Crop/MT/PS treatment. These results demonstrate that the way soil is managed influences its suppressiveness to Pythium rhizome rot. It should therefore be possible to reduce the impact of this disease by modifying the current ginger farming system.

Table. Impact of rotation, tillage and amendment treatments on suppressiveness to Pythium rhizome rot

Rotation/tillage/ Amendment	Disease Severity[AB]	% Plants Remaining Symptomless[B]	
Kandanga		IR	NIR
Fumigated/CT/Nil	2.65 b	0	5
Crop/CT/Nil	2.25 ab	6	16
Crop/CT/PS	2.05 ab	9	18
Crop/MT/Nil	2.35 b	17	5
Crop/MT/PS	1.68 a	5	24
Pasture/Nil	2.60 b	0	0
Pasture/PS	2.41 b	5	16
Yandina			
Fallow/CT/Nil	1.93 b	9	25
Crop/CT/Nil	1.607 ab	5	35
Crop/CT/PS	1.39 ab	22	33
Crop/MT/Nil	1.44 ab	11	42
Crop/MT/PS	0.81 a	0	65
Pasture/Nil	1.61 ab	17	40
Pasture/PS	0.89 ab	20	55

Note:

For each site, numbers in each column followed by the same letter are not significantly different (P = 0.05). ADisease severity was assessed in non-irradiated soils after 3 months.

BAssessments were based on 17-21 plants.

Plant Parasitic Nematodes on Ginger

To develop a clearer understanding of the plant parasitic nematode populations in the ginger crop it was decided to initiate a more detailed study across farms, districts and crops used in rotation with ginger. Populations were determined at (or near) the time of planting (P¬i) and at (or near) the time of crop harvest (Pf). Initial and final nematode population densities (P¬i and Pf, respectively) were determined in typical ginger-growing fields in the Navua, Veikoba and Waibau regions. Ginger and the crops that follow it in the rotation (cassava and taro) were sampled at the time each crop was planted and harvested. Nematodes were extracted by spreading 200 ml soil on an extraction tray. Multiplication rates for particular nematodes were determined as Pf/P¬i.

Multiplication rates for R. similis on taro and cassava were determined in 1 L pots of pasteurised sand and peat. Pots were inoculated with 1,000 R. similis and 40 weeks later, nematodes were extracted from soil and from roots that were macerated in a blender and then spread on an extraction tray.

R. reniformis was the most common plant-parasitic nematode in ginger-growing soils. It multiplied readily on ginger, whereas there was little multiplication on taro or cassava.

Table. Initial and final populations (Pi and Pf), and multiplication rates (Pf/Pi) of reniform nematode (Rotylenchulus reniformis) on ginger, taro and cassava in various fields in Fiji

Crop	Nematodes/200 mL soil		
	P_i	P_f	P_f/P_i
Immature ginger	93	2410	25.9
	30	710	23.7
	540	7600	14.1
Mature ginger	805	3550	4.4
	370	4560	12.3
	40	830	20.8
Taro	1535	350	0.2
	90	104	1.2
	2440	1328	0.5
	1640	1532	0.9
	530	672	1.3
Cassava	390	280	0.7
	436	130	0.3
	490	380	0.4
	1120	1560	1.4
	1010	380	0.4

Root-knot nematode was not present in enough fields to allow definite conclusions to be drawn about its multiplication rates on each crop. Also, final population densities on ginger were rarely greater than 200 nematodes/200 ml soil, while populations on taro and cassava were even lower. Occurrence was dependent on soil texture, with the nematode most common in well-structured clay loam soils of Waibau and silt loam soils of Navua.

R. similis occurred on a few farms in the Veikoba region. The nematode was often not detectable in soil at the time ginger was planted, but was invariably introduced in planting material. It then multiplied rapidly and was readily detected later in the season.

The highest population densities recorded from ginger fields were 170 nematodes/200 ml soil at immature harvest and 410 nematodes/200 ml soil at mature harvest. R. similis was never found in taro or cassava roots, but was occasionally found at low population densities in soil associated with roots of those crops.

However, moderate to high populations were invariably found on volunteer ginger, which was always present as a weed in the following taro and cassava crops. Results from the pot test indicated that taro was not a host of R. similis. The nematode was recovered from cassava roots, but the final nematode population density was lower than the number of nematodes inoculated.

Table. Final nematode population densities on five replicate plants of taro and cassava 40 weeks after they were inoculated with 1,000 Radopholus similis

Crop	No. R. similis		
	Roots	/200 mL soil	/pot
Taro	0	0	0
Cassava	34 ± 17	64 ± 28	379 ± 168

In the Fijian ginger farming system, ginger is generally grown every 3 years, with crops of cassava and taro planted in the years between ginger crops. Our results indicate that this is an excellent rotation for managing the three plant-parasitic nematodes likely to cause damage on ginger. Taro and cassava are poor hosts of both R. reniformis and R. similis, while Meloidogyne spp. rarely reached high population densities in ginger-growing soils, regardless of the crop that is grown.

Through the survey work, a farm was identified in the Veikoba district to conduct more detailed field studies with the burrowing nematode (R. similis), the most destructive of the plant parasitic nematodes of ginger in Fiji.

Our pathogenicity experiments have shown that R. similis multiplies initially on ginger roots and seed pieces and then invades newly-developing rhizomes.

The nematode seems to feed first on outer parts of the rhizome and the resulting damage leads to yellowing at the base of shoots and of the lower leaf sheath. As more tissues are destroyed, older leaves turn yellow, shoots eventually collapse and discolouration extends further into the rhizome. The end result is that plants eventually die and the rhizome is totally destroyed. Given the results of our experiments, we suggest that R. similis is a pathogen of ginger in its own right. The nematode is capable of killing plants and destroying rhizomes, with secondary organisms playing little role in symptom development.

Field results at Veikoba mirrored those obtained in pots. They indicated that Radopholus multiplies readily on ginger but when initial nematode population densities are low, nematodes generally do not multiply to levels capable of causing economic losses to immature ginger.

Also nematode numbers can vary markedly from plant to plant and this is correlated with levels of infestation in seed used for planting. Other findings were that nematode populations can be maintained between ginger crops on volunteer ginger and on a number of common weeds. These infestations can act as 'hot spots' for further spread when ginger is replanted. However the finding that Radopholus was found on the farmer's seed is of most concern as it demonstrates unequivocally that Radopholus and other parasitic nematodes are being introduced on planting material.

Continued observations at the Veikoba farm demonstrated that the rotation with taro and cassava, together with the preparation of the beds with poultry manure, had significantly reduced parasitic nematode populations to the point that the two most serious plant parasitic nematodes of ginger, burrowing nematode and root-knot nematode, were undetected in the beds to be planted to ginger.

However they have been detected in seed already planted or about to be planted. Even though the grower was culling infested seed in his seed sorting area, it was estimated that approximately 30 per cent of the seed was infested and, in some instances, the infestation was very severe.

The results showed that populations of free-living nematodes increased to very high levels when poultry manure was added to the soil, largely because of an exponential increase in numbers of bacterial-feeding rhabditids. In contrast, populations of predatory mononchids were markedly reduced by even the lowest rate of poultry manure. Populations of R. reniformis and R. similis were significantly reduced by poultry manure at 20 t/ha.

Table. Populations of plant-parasitic and free-living nematodes six weeks after applying various rates of poultry manure to a ginger-growing soil

Poultry manure (t/ha)	Rotylenchulus reniformis	Radopholus similis	Mononchidae	Total free-living nematodes
0	227 a	37 a	71 a	3,630 b
10	189 a	34 a	1 b	27,541 a
15	152 a	33 a	0 b	63,095 a
20	41 b	4 b	0 b	85,112 a

Numbers in the same column followed by the same letter are not significantly different (P = 0.05)

Since the moisture content of the poultry manure was about 5 per cent and it contained 3.89 per cent N, the equivalent of 0, 370, 555 and 740 kg N/ha was applied in the various treatments. Given that much of this N would have been converted to ammonia and it is known to be toxic to nematodes at the concentrations likely to have been achieved with the highest application rate (Stirling 1991), it is not surprising that populations of plant-parasitic nematodes were reduced by the amendment. Although this might be seen by ginger growers as one of the benefits of applying poultry manure, its negative impact on the predatory nematodes that regulate nematode populations must also be considered.

Predatory nematodes are particularly susceptible to N inputs (Tenuta and Ferris, 2004) and their demise may explain why populations of plant-parasites resurge strongly when ginger is planted. Previous work on hot water treatment for nematode control in ginger was done with rhizomes infested with root-knot nematode (Vilsoni et al. 1981). Another experiment was established during this project to check that the recommended treatment is effective against burrowing nematode (R. similis).

In October 2009, ginger rhizomes infested with R. similis were obtained from Lee Tong's farm at Veikoba and either immersed in hot water (51°C for 10 minutes) or left untreated. They were then planted individually in 2L pots filled with pasteurised soil. Observations in February 2010 suggested plants in the hot water treatment were growing better than untreated plants. By late April 2010 the untreated seed were showing symptoms of damage caused by R. similis, however no nematodes were detected in the soil or roots of hot-water treated seed.

FARMING SYSTEMS FOR MANAGING SOIL-BORNE PATHOGENS

Edible ginger (Zingiber officinale) production in Fiji dates back to the 1950s when a small area was grown in the Suva peninsula with produce exported to New Zealand. In the 1960s the production dramatically increased as Fiji started exporting fresh ginger to North America and new production areas were identified.

Production increased to 5,000 metric tonnes in the 1980s and stayed around that figure for some time until 1999 when the North American market collapsed due to competition from China. The industry has since recovered and restructured with an emphasis on exports of fresh immature ginger and the entry of ginger processors that preserve immature ginger in brine or syrup for export markets. In exports alone, the industry was valued at FJD$4.2 million in 2004 when this project first commenced activities.

Ginger is planted in Fiji between September and early November, and immature (green) ginger is harvested January-March (70 per cent of the crop) and mature (brown) ginger in June-July. Ginger is also harvested for 'seed' which provides the vegetative planting material for next season's crop. Ginger is mainly grown in the wetter areas of Viti Levu (Suva/ Nausori/Navua) and predominantly by Fijian smallholders as an important cash crop. Exporters and processors contract growers to produce an agreed quantity of good quality ginger at a set price.

The past 15 years has seen a decline in ginger production and greater wastage due to soil-borne diseases mainly occurring during the wetter, warmer months. Several recent surveys, HortResearch/Fiji Government (1996-98) in Fiji and Biological Crop Protection/Horticulture Australia (1997-2001) in Australia, provide convincing evidence that soil-borne diseases are the major problem. A major constraint to production in Fiji is the soil-borne pathogen, Pythium myriotylum (Pm), which infects seed-pieces and causes rot and early death of plants.

Pm was killing up to 30 per cent of plants during surveys undertaken during 1997/98 but is capable of completely destroying the entire crop. In Australia the major soil-borne pathogen of edible ginger was widely regarded as Fusarium oxysporum f. sp. zingiberi (Foz), until the wet summer of 2007/08 when Pythium myriotylum was isolated on a few farms. Both are now responsible for extensive rhizome rots.

The epidemiology of both Pm and Foz is also influenced by nematode damage to roots and rhizomes, which can exacerbate disease development. Annual farm-gate losses due to soil-borne pathogens were estimated in 2004 at FJD$530,000 for Fiji and AUD$990,000 for Australia, and the situation was deteriorating.

These pathogens limit production and affect rhizome quality, which in turn can decrease profits and destabilise the market. Furthermore pressure is placed on 'seed' beds as a source of material to satisfy contracts. If the best material is sold and pathogen infected material used for planting, then a downward spiral begins. If serious enough, pathogens can even lead to market failure. It was important that the disease/pest cycle be broken.

A review of the ginger industry was undertaken by the Fijian Ministry of Agriculture in conjunction with a major growers' meeting during 2004.

The major issues and priorities, as listed at the meeting, included:

- Increasing threat of disease – production losses/poor quality/lack of seed ginger
- Increasing cost of production under current farming system
- Scarcity of labour
- Low labour productivity
- Lack of credit facilities
- Environment degradation
- Urban drift – farming no longer an attractive option for the younger generation
- Institutional problems

Despite these concerns the industry concluded, "Fiji ginger is known for its taste but can no longer be complacent in terms of quality and price. There is great potential for expanding the processing sector and fresh immature ginger may become a specialty product in the hands of a few exporters. Fiji is capable of producing quality ginger and at a competitive price providing the problems highlighted above are addressed effectively and in time." It is interesting to note that the priorities identified by the Fijian ginger industry are almost identical to the Australian ginger industry, and in the same order.

ACIAR project PC/2004/049 attempted to address the priority issue of soil-borne diseases and the use of more sustainable production practices should also address the issue of environmental degradation. Creation of a more profitable ginger industry should have flow on benefits for regional communities as well as businesses in the larger towns.

RESEARCH STRATEGY

The research strategy for the project involved developing a farming systems approach to ginger production using pathogen-tested planting material, bedding/mounding, reduced tillage, crop rotation and the use of mulches/organic amendments to improve soil physical and chemical properties and suppressiveness to soil-borne pathogens. The key was to achieve an active soil biology that was antagonistic to fungal pathogens, enhance populations of free-living nematodes to levels that maintain predators of nematodes (particularly fungal predators), and limit nitrogen to levels that encourage predatory activity (Stirling 1989; 2004).

Improvements in soil health status were confirmed with some simple measurements of microbial activity, free-living nematodes, as well as measures of nutrient status and soil organic matter. Combining this with yield data and disease incidence enabled us to determine whether we improved the soil's suppressiveness to key soil-borne pathogens and restored its biology. This knowledge immediately translates into practical recommendations for the most appropriate farming systems to use for ginger production in Fiji and Australia.

This project drew on findings related to ginger marketing and economics of the ACIAR project ASEM 2003/069 Policy options for improving the value of land use in smallholder Fijian Agriculture. This project undertook a small study to look at the economics and marketing of ginger during the second year of its implementation through a Masters Research project by Waisiki Gonemaituba at the University of Queensland. This project also had links with other ACIAR projects in the Pacific dealing with IPM of vegetables *e.g.* CP 2004/063 Integrated pest management in a sustainable production system for Brassica crops in Fiji and Samoa.

A farmer participatory research (FPR) approach was used at 3 farmer trial sites in Fiji (Waibau, Navua and Veikoba) and 2 trial sites in Australia (Kandanga and North Arm). This research was also linked with the Fijian Ministry's extension service and a senior officer was devoted full-time to ginger and six other officers were located in the major ginger producing areas with 30 per cent of their time responsible for providing support to ginger growers in the area. In addition the exporters/processors, who have a vested interest in the supply of quality material, have liaison officers who work with farmers as well as the Ministry's Extension Division. These people knew the growers and had the necessary social-cultural skills to help drive the FPR model. FPR, both in Fiji and Australia, had the advantage that industry champions were identified so the project was tailored to more effectively deliver agreed outputs.

The Fijian government targeted support for export crops as very high priority and there was strong support and commitment for the ginger project at all levels. In this context the Ministry had a policy in place to set up model farms to not only provide demonstrations of best farming practice, but to provide industry support in the form of clean planting material.

Develop Practical Systems for Production of Clean Planting Material

The methods for tissue culture propagation of ginger are well established (Smith and Hamill, 1996) and in routine use at Maroochy Research Station and Koronivia Research Station. Small trial plots of tissue cultured material were established in Fiji and Australia and seed-pieces harvested from these plots were used to establish nursery beds for the continued production and supply of superior quality 'seed' material.

Nursery sites were identified that have not grown ginger or used water from dams for past ginger production. The nursery sites were also managed as if they were quarantined from the rest of the ginger industry. Sustainability of such schemes will depend on clear demonstration of benefits and results from our studies have highlighted the value of clean 'seed' as compared to the 'seed' currently in use. This is particularly important in Fiji to stop the practice of buying second-grade seed or in using market reject material as planting material.

Seed treatments consisted of allowing the seed to suberise for 4-5 days under ambient temperatures vs seed that had been cut and immediately planted, as well as hot-water treatments where seed was dipped in water maintained at either 48°C for 20 minutes or 54°C for 15 minutes.

DEVELOP SOIL MANAGEMENT SYSTEMS

Observations on Crop Losses and Disease Etiology

Fijian fields were checked for signs of Pythium Soft Rot and plants with yellow or dead shoots were inspected. If symptoms were typical of soft rot, rhizomes were collected and the diagnosis confirmed by isolating the pathogen in the laboratory. Ginger brought to the processing factory in February 2007 was also checked for soft rot and other pests and diseases. Twenty consignments of ginger from various parts of Fiji were carefully inspected, the amount rejected for various reasons was recorded and pathogens associated with disease symptoms were confirmed in the laboratory. Since most Fijian crops are harvested in February as immature ginger, observations on the development of rhizome rot later in the season were limited to crops retained for planting material.

The spatial distribution of the disease was monitored in eight fields (four near Veikoba and four in the Waibau region) that ranged in area from 48 to 240 m2. All crops were disease-free in February 2008 and were therefore retained to provide seed later in the season. Fields were inspected in March, April and May 2008 and the location of all plants showing disease symptoms was marked on a map. Final observations were made when crops were harvested for seed during late July and early August 2008 and seed losses were estimated in consultation with the grower.

Pythium Soft Rot

Pythium Soft Rot had never been observed on ginger in Australia until late December 2007, when a severe outbreak occurred on two farms in Queensland that were 28 km apart but owned by the same family. Initial observations were made soon after the disease was reported by the grower and disease spread was then monitored in four fields that had distinct patches of diseased ginger surrounded by green, healthy plants. Three of the fields were at Eumundi and one at Kandanga, and they varied in size from 1.4 to 3.6 ha. Datum plots (10 m long with three rows of ginger contiguous along beds 1.9 m wide) were marked out in each field and disease severity was rated every few days as follows:

0 = no yellow and necrotic shoots in the datum plot;
1 = <2%;
2 = 3–5%;
3 = 6–10%;
4 = 11–33%;
5 = 34–67%;
6 = 68–100% diseased shoots in the datum plot.

The rate of disease spread was measured by first marking the edge separating diseased and yellowing shoots from green and healthy shoots. The daily rate of spread was calculated by measuring the distance the diseased edge had moved between recording dates. Fields were monitored every few days from mid-January to early February 2008, and final observations were made just before harvest of immature ginger in late February 2008.

Rainfall and Temperature Data

Since outbreaks of soft rot are known to be associated with hot and consistently wet conditions, temperature and rainfall data were collated for two sites: Koronivia Research Station, Nausori, Fiji and Maroochy Research

Station, Nambour, Queensland, Australia. Both stations are located within 20 km of the major ginger-growing areas in each country and have a climate that is broadly similar to those areas. The records date back to 1973 for Nausori and 1952 for Nambour. Long-term averages for December–May and data from the 2006–07 and 2007–08 ginger-growing seasons were retrieved for rainfall, the number of rain days and minimum and maximum temperature.

Isolation and Identification of Pythium

Ginger rhizomes showing disease symptoms were washed thoroughly in water, dipped in ethanol and flamed briefly before tissue from the leading edge of lesions was transferred to media that were selective for Pythium. In Fiji, isolations were done on cornmeal agar with carbendazim (2 mg/mL), ampicillin (250 mg/mL), rifampicin (10 mg/mL), pimaricin (5 mg/mL), and pentachloronitrobenzene (100 mg/mL), while a medium containing penicillin, polymixin and pimaricin was used in Australia (3P medium of Eckert and Tsao 1962). To check for fungal pathogens other than Pythium, tissue was also plated onto potato dextrose agar containing streptomycin (120 mg/mL).

Pythium isolates were identified using methods outlined by Van der Plaats-Niterink (1981). Growth rates at various temperatures were determined on potato carrot agar while sporangia, antheridia, oogonia and oospores were produced by transferring the fungus to autoclaved grass leaves floating in sterile water. The identity of isolates used in pathogenicity tests was confirmed by sequence analysis of rDNA internal transcribed spacer (ITS) regions containing ITS1 and ITS2 and intervening 5.8rDNA using the primer pair ITS1/ ITS4 (White et al. 1990).

Pythium Soft Rot Pathogenicity Experiments

Experiments were set up in Fiji and Australia using isolates of P. myriotylum from each country. Inoculum was prepared by placing 50 g sorghum seed in 250-mL Ehrlenmeyer flasks, soaking the seed overnight in water, pouring off the excess water, autoclaving twice on successive days and then inoculating the seed from cultures on agar.

The inoculum was used after the oomycete had fully colonised the substrate (usually 7–10 days at 25°C). Plants were grown in pots containing steam-sterilised potting mix (3 sand: 1 peat) and were inoculated 2–5 months after a seed piece had been planted in each pot. After 4 g of inoculum was added, the soil was kept moist for 4 days and then a wet/moist cycle (4 days wet and 3 days moist) was imposed by placing a saucer under the pot for 4 days and removing it for 3 days.

At the end of each experiment, small pieces of tissue from rotting rhizomes were transferred onto a selective medium to check for the presence of Pythium. In both experiments, the level of disease in each pot was assessed periodically by rating aboveground symptoms using the following scale: 0 = healthy plant;

1 = some leaves starting to turn yellow; 2 = plant alive, but shoots either totally yellow or dead; and 3 = all shoots dead.

Plant Parasitic Nematodes

Surveys of plant parasitic nematodes that could also be responsible for yield losses were conducted in farmers' fields from all the major ginger growing areas in Fiji. Soil was collected from around the roots and rhizomes, mixed gently and a 200 ml sample was placed on a standard extraction pan for 2 days. Nematodes recovered by sieving twice on a 53 μm sieve were counted.

A total of 20 consignments of ginger were checked at the Frespak factory at Lami by inspecting ginger on the conveyor belt after washing and in the reject bins. About 100 rhizome pieces in each consignment were carefully inspected and lesions possibly caused by nematodes were teased out under water to check for the presence of root-knot nematode (Meloidogyne spp.) and burrowing nematode (Radopholus similis). The washed weight of each consignment and the weight of ginger rejected were obtained from factory records.

Radopholus. Relatively small (5-15 mm diameter), light brown or water-soaked lesions generally extending only 1-2 mm below the surface. Meloidogyne. Slight eruptions of tissue usually near the base of developing buds.

Burrowing Nematode Pathogenicity Experiment

On 12 September 2008, a potting medium consisting of 2 parts washed river sand and 1 part commercial potting mix was heated to 70ºC for 30 minutes. Twenty 4L pots were then filled with this medium and a metalaxyl-treated seed piece of ginger (Z. officinale) was planted in each pot. Pots were transferred to a glasshouse and after 6 weeks, when plants had shoots up to 10 cm long, half the pots were inoculated with 1,500 Radopholus similis. The nematode was originally obtained from a ginger farm at Veikoba, Fiji and had been multiplied in the laboratory on sterile carrot tissue.

On 9 February and 20 March 2009 (15 and 20 weeks after pots were inoculated), the total number of shoots and the number of yellowing or dead shoots in each pot was recorded. At each harvest date, the above-ground biomass in five inoculated and five control pots was measured and seed pieces and newly-developing rhizomes were then removed from the pots for assessment. Roots were chopped into segments <1 cm long and mixed with the soil, and a 200 ml sample (about 220 g) of soil and roots was spread on a Baermann tray. After 2 days, nematodes were recovered on a 38 μm sieve.

Portions of seed pieces and rhizomes that showed symptoms possibly caused by R. similis were assessed by removing small pieces of tissue (about 1 mm3) from affected areas, macerating them in water and checking for nematodes after 24 hours. Discoloured tissue was also checked for fungal pathogens by placing small pieces of the tissue onto potato dextrose agar (PDA) or corn meal agar with carbendazim, ampicillin, rifampicin, pentachloronitrobenzene and pimaricin (CARPP) and observing plates after 24 and 48 hours.

The nematodes present in larger pieces of discoloured or rotting tissue (*i.e.* 0.5 – 4 g) were recovered by macerating the tissue in 200 ml of water in a blender and then placing it on a Baermann tray for 24 hours. This maceration technique was also used to recover nematodes from the remaining part of each seed piece or rhizome. In cases where nematodes were extracted from several pieces of tissue from the same seed piece or rhizome, the counts were added to obtain the total number of nematodes.

Developing Farming Systems to Control Soil-borne Pathogens

In Fiji a field trial at Waibau was established in September 2006 to examine the effect of nitrogen (N) on Pythium Soft Rot. The experiment consisted of four rates of poultry manure (0, 5, 12.5 and 17.5 t/ha) applied before planting to four replicate plots (each 20 m2) set out in a randomised block design. In Australia the trials were larger, with more treatments and larger blocks to accommodate mechanised operations. An experiment was established to examine the influence of cover cropping, pasture leys, organic amendments and minimum tillage in a ginger farming system over 2 crop cycles.

Soil textural and chemical analyses were made from soil sampled to a depth of 10 cm and major and minor elements were measured using a commercial nutrient analysis service in Australia (Incitec Pivot, Werribee) or the Soil Testing Laboratory at Koronivia Research Station.

The presence of soil-borne pathogens was confirmed by collecting soil samples from a depth of 0-10 cm on 6 April 2006. The soil was mixed gently, spread on a standard nematode extraction tray and after 2 days, nematodes were recovered by sieving twice over a 38 μm sieve. Bioassays in which tissue-cultured ginger plants were planted into 1 L of field soil diluted either 1/10 or 1/100 with autoclaved soil and grown in a glasshouse for up to 8 months to detect presence of fungal pathogens.

The basic experiment consisted of 5 cropping/tillage treatments × ± poultry manure amendment × 4 replicates. The cropping tillage treatments were set out in a randomised block design, with the amendment as a split plot.

The main treatments were:

- Conventional tillage with continuous cropping and a conventional tillage ginger crop,
- Minimum tillage with continuous cropping and a conventional tillage ginger crop,
- Conventional tillage with a summer crop only and a conventional tillage ginger crop,

- Minimum tillage with a summer crop only and a minimum tillage ginger crop, and a four-year pasture ley further split into (ECT) conventional tillage ginger crop and (EMT) minimum tillage ginger crop. A fumigated control (F) was included where conventional tillage summer crops were followed with a conventional tillage ginger crop, however the beds were treated with 500 kg/ha 1,3-dichloropropene (Telone®) 3-4 weeks prior to planting ginger. Field experiments were grown as per commercial practice.

Growth and Yield Measurements of Ginger Crop

Shoot emergence and number of shoots were counted from a 3 m length of bed on the 26 October 2009, 11 November 2009 and 3 December 2009. Rhizomes were harvested from a 1 m section of bed and an estimate of yield extrapolated from the weight of uninfected rhizomes (t/ha). Rhizome that had been rejected due to rot was weighed and expressed as a percentage of the total rhizome harvested from the plot.

The first-late harvest occurred on 4 May 2010 and again rhizomes were harvested from a 1 m section of bed and both rejected (due to rot) and sound rhizomes were weighed and yields calculated. Root knot nematode damage was slight (none would have been rejected) and was expressed as the percentage of 10 randomly selected rhizomes that showed any signs of damage.

Soil-borne Pathogen Assessments

For the first ginger crop at early harvest on the 12 March 2008, the ginger in a datum plot 1.15 m long was harvested and the number of rhizomes showing rot symptoms caused by soilborne fungi was recorded. Laboratory isolations indicated that these symptoms were primarily caused by Pythium myriotylum. However by mid-April yellowing of plants was associated with Foz infection of the rhizomes and it was estimated that another 10 per cent of the trial was affected. On 14 April 2008 (2 days prior to first-late harvest), three independent observers estimated the percentage dead and severely yellowing shoots in an 18 m datum plot in the centre of each plot. Estimates were then converted to a disease rating of 1 to 6, where a rating of 1 indicating that <5 per cent of shoots showed symptoms and a rating of 6 indicating 45-55 per cent of shoots were affected.

For the second ginger crop and prior to early harvest (31 March 2010), the percentage of yellowing and necrotic shoots in the entire length of the bed was estimated for each treatment, and plots were rated for disease severity (0 = all shoots green and healthy to 5 = total collapse and death of shoots).

Root knot nematode populations were monitored during the growing season by collecting soil samples prior to planting, at early harvest and 2 days prior to first-late harvest by the method described previously. At each sampling time, 10 cores were collected from each plot to a depth of 0-20 cm using a 2.5 cm diameter sampling tube. Since root knot nematode was not detected in soil samples collected prior to planting of the first ginger crop, the nematode was inoculated at one point in each plot in an attempt to determine whether some treatments were more suppressive to nematode multiplication than others.

Inoculum of M. javanica was obtained by retrieving eggs from glasshouse-grown tomato plants using NaOCl and 50,000 nematode eggs were then inoculated at a depth of 15 cm, at a point 4.5 m from the end of each plot. Eggs were added on 15 October 2007, three weeks after ginger was planted but before any emergence had occurred.

On 3 March 2008 (early harvest), all plants within 20cm of the inoculation point were dug up and all rhizomes (generally 3-5 pieces) were rated for root knot nematode damage on the following scale: 0 = no damage; 1 = a low level of damage (1-2 areas erupted due to the nematode per rhizome); 2 = moderate damage (3-5 erupted areas); 3 = high (6-10 erupted areas) and 4 = very high (rhizomes unmarketable due to root knot nematode damage). After rhizomes had been removed, the soil in this area was sampled and nematodes were extracted and counted as described previously.

Pythium Suppression Bioassays

In September 2009 (just before ginger was planted and four years after the experiment commenced), soil was collected from each plot and half was irradiated at 25 kGy. Both irradiated (IR) and non-irradiated soils were placed in 2 L plastic planter bags and planted with a 60 g piece of ginger rhizome. Pythium myriotylum was cultured on autoclaved sorghum seed and on 10 December 2009, when plants had 1 or 2 shoots up to 50 cm long, each pot was inoculated with two Pythium-colonised seeds.

The assay was repeated on 19 January 2010, except each pot received only one colonised seed. The mean maximum temperature during December and January was over 30°C and overhead sprinklers in the screen house provided a watering regime conducive for Pythium infection. After approximately 3 months for each assay, plants were assessed for disease severity on a 0-3 scale where 0 = no disease; 1 = some leaf yellowing; 2 = most shoots yellow or dead; 3 = rhizomes rotted and plants dead.

Assessments were based on 17-21 plants for each treatment in the first assay and 11-16 plants per treatment in the second assay. Pythium was confirmed in a random selection of affected rhizomes at the end of the assays.

ON-FARM IMPLEMENTATION OF IMPROVED GINGER PRODUCTION SYSTEMS

Farms for farming systems test sites were selected in consultation with researchers, extension staff, as well as leading exporters/processors. Three sites were chosen to conduct trial work in Fiji, while there were two in Australia.

Key staff met with selected growers before crop planting to discuss pathogen and crop problems and their experience with different practices, to prioritise problems and possible alternatives for testing, and to identify indicators for evaluating progress. At key periods in the crop cycle, they met again to compare plant growth and disease levels. They also discuss the progress of their experiments. At each meeting they also conducted field visits and discovery exercises to strengthen their ecological knowledge of the crop, pathogens and soil factors. At the end of the cycle, they identified major lessons to be learned in the next crop cycle and problems for further experimentation or study.

Provision of Tissue Cultured Plants and Clean Seed

A clean-seed scheme operated in Fiji in the past and there is interest from a number of exporters to reintroduce the scheme. Koronivia Research Station also has a tissue culture laboratory that is currently producing tissue cultured ginger for agronomic evaluation. An inspection of the facility and discussions with staff revealed that capacity exists within the Ministry to multiply and harden-off tissue cultured ginger plants.

In Fiji, two sites have been established by the Ministry of Primary Industries for the production of clean seed for use by industry: one is in the Western Division of Viti Levu near Nadi and the other in the Northern Division of Vanua Levu at Seaqaqa Research Station. These areas are declared free of Radopholus similis and helped alleviate short-falls experienced in supply of planting material during the 2008/09, 2009/10 and 2010/11 seasons.

Tissue cultured plants of the improved ginger variety, 'Buderim Gold', were imported to Fiji under permit and established at Koronivia Research Station (KRS). 'Buderim Gold' is a tetraploid white ginger cultivar developed by the former Queensland Department of Primary Industry and Fisheries (QDPI&F) and has significantly larger rhizomes compared to the mother plant, 'Queensland'. After discussions with Fijian fresh ginger exporters on previous visits, they felt a market may exist for this product in Japan and the USA. After discussions with Buderim Ginger Limited, who helped fund the developmental work on 'Buderim Gold', it was decided to release it in Fiji and determine if international markets may exist for a large, white fresh ginger product.

Most of the 'Buderim Gold' plants were established in pots in a glasshouse under shade, but six plantlets were retained for tissue culture multiplication at KRS, while another six plantlets were to be placed in the in vitro germplasm collection at the Secretariat of the Pacific Community (SPC).

Two Australian 'seed' producers were each supplied with consignments of 1500 tissue cultured plants over the course of the project. This was part of an on-going process to ensure that disease-free and pest-free material is introduced into the nursery production system. In addition, the project team continued to test for soil-borne pathogens on these farms to ensure the spread of diseases and pests on planting material is minimised. Demand for clean seed by the ginger industry continues to exceed supply.

Seed Treatments

In Fiji, seed is typically allowed to air-dry under shade for 8-10 days before planting, whereas the economic imperative in Australia is to plant seed as soon as possible following cutting and fungicide treatment. This means that cut surfaces may not be sufficiently suberised before planting and the physical barrier to prevent pathogen invasion may not be fully formed. Pot experiments were conducted in the glasshouse to test the effect of seed drying treatments on germination and susceptibility to pathogens.

When planted into sterilised soil, seed germination was faster when not allowed to dry for 10 days, even though 17 per cent developed Erwinia soft rots in soils that were occasionally watered and this increased to 35 per cent when soils where maintained at field-capacity. Seed that had been allowed to dry for 10 days and had a suberised barrier at the cut end did not develop any rots. These experiments suggest that seed requires a period for suberisation to occur on the cut ends prior to planting to minimize rots and improve establishment of the ginger crop.

In Australia, seed is hot-water treated at 48oC for 20 min and then dipped in fungicide to control Meloidogyne. This is often restricted to 'seed' being used to establish nursery blocks as the hot-water treatment can reduce yields by up to 20 per cent. This is deemed acceptable because the rhizomes used for next season's planting material must remain undisturbed in beds for up to 12 months.

In Fiji, A small (1.2 m3) gas-fired vat was loaned by the Ministry of Agriculture for the purposes of hot-water treating ginger seed. A farmer at the Veikoba trial site used his 'standard' method of treating seed which consisted of half-filling the vat with approximately 600 L of water, waiting until the temperature of the water approached 50oC (varied from 48oC to 50oC) and placing a maximum of eight 20 kg bags of seed in the vat and leaving for 15 min. The temperature of the water was 40oC at the end of this 'standard' treatment. Our project team was allowed to treat one bag of the grower's own seed using the Ministry's recommended treatment of 54oC for 10 min (this time the temperature was held constant over the treatment period). A sample of seed was collected and tested. Whereas live nematodes were recovered from seed using the farmer's 'standard' treatment, no live nematodes were detected using MAF's recommended procedure. A significant finding was that Radopholus was detected in grower's seed from 4 of the 8 plots sampled. This is strong evidence that burrowing nematodes are being introduced into prepared plots on seed that has been insufficiently hot-water treated.

Pest and Disease Surveys

The evidence collected from Fiji and Australia strongly supports the association of Pythium Soft Rot with long periods of wet, overcast weather and in situations where the soils remain saturated.

All Pythium isolates obtained from ginger in Fiji and Australia grow well at high temperatures. They continue to grow slowly at 41oC and have radial growth rates on agar of about 30 mm/day at 37oC. The Australian isolates from ginger have been identified as P. myriotylum and possibly P. zingiberis, whereas the Fijian isolates have been identified as P. myriotylum, P. graminicola and P. vexans. Pathogenicity tests performed in Fiji and Australia confirmed we are dealing with highly pathogenic isolates that are more virulent when soils are saturated and infection does not required damage to the host roots or rhizomes as caused by the nematodes Radopholus or Meloidogyne. It was established that Radopholus is a primary pest causing damage to the roots and rhizome, but economic losses are negligible for the immature harvest. However by the harvest of mature ginger, particularly for rhizome that is used as 'seed', damage can be severe and the wound sites lead to development of rots that can result in extensive losses in the field.

KEY FEATURES OF SOILBORNE PLANT PATHOGENS

Disease Symptoms

Many soilborne fungal plant pathogens cause disease of the roots or stem disrupting the uptake and translocation of water and nutrients from the soil. Therefore they commonly cause similar symptoms to drought and nutrient deficiencies; these include wilting, yellowing, stunting and plant death. Thus symptoms such as this do not necessarily indicate that a pathogen is causing disease.

Host Range

The majority of these pathogens have a wide host range and may cause different types of diseases on different hosts. A particular species may cause severe root rot on one host but only cause superficial and symptomless infection on the roots of another host. Crop rotation is not necessarily an appropriate control measure for some of these pathogens because of their wide host range which may include weeds. However some pathogens have a very narrow host range, *e.g.* forma speciales of the wilt pathogen Fusarium oxysporum. However, many of these formae speciales can persist by symptomless colonisation of the cortex of roots of non-host plants.

Epidemiology and Survival

In general these fungi can persist for long periods in soil in the absence of susceptible crops. Some species persist mainly as resistant hyphae in plant residues, in large pieces of stubble or small fragments. The inoculum of such pathogens declines gradually as the residues decompose. Decomposition is hastened by warm wet conditions and mechanical disruption caused by cultivation. Many species also persist as propagules which are adapted for long-term survival in soil. These include thick walled spores such as oospores and chlamydospores, together with sclerotia and microsclerotia.

ENVIRONMENT OF DISEASES CAUSED BY SOILBORNE PLANT PATHOGENS

Root Rot

These diseases are caused by a diverse group of fungi and related organisms. The most important genera include Pythium and Phytophthora, Rhizoctonia, Cylindrocladium and Armillaria.

These diseases are characterised by a decay of the true root system; some pathogens are generally confined to the juvenile roots whilst others are capable of attacking older parts of the root system. Symptoms that are observable include wilting, leaf death and leaf fall, death of branches and limbs and in severe cases death of the whole plant. Some examples of these diseases are shown below.

Stem, Collar and Head Rots

These diseases are also caused by a diverse group of pathogens including species of Phytophthora, Sclerotium, Rhizoctonia, Sclerotinia, Fusarium and occasionally Aspergillus niger. The most obvious symptom of these diseases is the decay of the stem at ground level. Quite often this decay can lead to symptoms of wilting, death of leaves and to death of the plant. Some examples of these diseases are shown below. In tropical agricultural ecosystems these fungi can cause stem, foliage and head rots during warm, wet conditions. Phytophthora spp. for example, can cause diseases such as heart rot of pineapple, blight of potato and tomato and some fruit rots in these conditions. Similarly Rhizoctonia spp. can cause leaf blight in maize and head rot of cabbage in warm wet weather.

Wilt Diseases

The main species of fungi that cause these diseases are Fusarium oxysporum and Verticillium spp. The symptoms of these diseases include wilting of the foliage and internal necrosis of the vascular tissue in the stem of the plant. Some species of bacteria can also cause similar types of diseases. Some examples of these diseases are shown below.

Seedling Blights and Damping-off Diseases

Various common names are used for diseases of seedlings such as seedling blight and damping-off. The fungi that commonly cause seedling death include Pythium, Phytophthora, Rhizoctonia, Sclerotium rolfsii and less commonly Fusarium spp.. These fungi can infect the seedling during the germination, pre-emergence or post-emergence phases of seedling establishment. Environmental factors which inhibit germination and emergence usually increase disease severity. Thus cold conditions, dry, or very wet soils or a hard soil surface commonly lead to increased seedling disease. In northern Vietnam Pythium, Rhizoctonia and Sclerotium rolfsii are commonly associated with seedling death of vegetables such as beans, cabbages and other cruciferous crops, cucurbits and tomato.

7

Agricultural Biotechnology and Food Microbiology

INTRODUCTION

Food microbiology is the study of the microorganisms that inhabit, create, or contaminate food. Of major importance is the study of microorganisms causing food spoilage. "Good" bacteria, however, such as probiotics, are becoming increasingly important in food science. In addition, microorganisms are essential for the production of foods such as cheese, yogurt, other fermented foods, bread, beer and wine.

The focus of Food Microbiology is on the detection and analysis of foodborne spoilage microorganisms. Food microbiology is the study of food micro-organisms; how we can identify and culture them, how they live, how some infect and cause disease and how we can make use of their activities.

Microbes are single-cell organisms so tiny that millions can fit into the eye of a needle. They are the oldest form of life on earth. Microbe fossils date back more than 3.5 billion years to a time when the Earth was covered with oceans that regularly reached the boiling point, hundreds of millions of years before dinosaurs roamed the earth.

The field of food microbiology is a very broad one, encompassing the study of microorganisms which have both beneficial and deleterious effects on the quality and safety of raw and processed foods. Food science is a discipline concerned with all aspects of food - beginning after harvesting, and ending with consumption by the consumer. It is considered one of the agricultural sciences, and it is a field which is entirely distinct from the field of nutrition. In the U.S., food science is typically studied at land-grant universities.

Examples of the activities of food scientists include the development of new food products, design of processes to produce these foods, choice of packaging materials, shelf-life studies, sensory evaluation of the product with potential consumers, microbiological and chemical testing. Food scientists in universities may study more fundamental phenomena that are directly linked to the production of a particular food product. Food scientists are generally not directly involved with the creation of genetically modified (bio-engineered) foods. Some of the subdisciplines of food science: Food safety, Food engineering, Product development, Sensory analysis, Food chemistry.

The primary tool of microbiologists is the ability to identify and quantitate food-borne microorganisms; however, the inherent inaccuracies in enumeration processess, and the natural variation found in all bacterial populations complicate the microbiologists job. Without microbes, we couldn't eat or breathe. Without us, they'd probably be just fine. Understanding microbes is vital to understanding the past and the future of ourselves and our planet.

Archaea look and act a lot like bacteria. So much so that until the late 1970s, scientists assumed they were a kind of "weird" bacteria. Then microbiologist Carl Woese devised an ingenious method of comparing genetic information showing that they could not rightly be called bacteria at all. Their genetic recipe is too different.

So different Woese decided they deserved their own special branch on the great family tree of life, a branch he dubbed the Archaea. Archaea comes from the Greek word meaning "ancient." An appropriate name, because many archaea thrive in conditions mimicking those found more than 3.5 billion years ago. Back then, the earth was still covered by oceans that regularly reached the boiling point — an extreme condition not unlike the hydrothermal vents and sulfuric waters where archaea are found today.

Some scientists consider archaea living fossils that may provide hints about what the earliest life forms on Earth were like, and how life evolved on our planet. In addition to superheated waters, archaea have been found in acid-laden streams around old mines, in frigid Antarctic ice and in the super-salty waters of the Dead Sea. A number of other extreme-living bacterial species also enjoy these conditions, too, such as the community of cyanobacteria and bacteria shown top right.

Foodborne illness or food poisoning is caused by consuming food contaminated with pathogenic bacteria, toxins, viruses, prions or parasites. Such contamination usually arises from improper handling, preparation or storage of food. Foodborne illness can also be caused by adding pesticides or medicines to food, or by accidentally consuming naturally poisonous substances like poisonous mushrooms or reef fish. Contact between food and pests, especially flies, rodents and cockroaches, is a further cause of contamination of food.

FOOD SAFETY

Food safety is a major focus of food microbiology. Pathogenic bacteria, viruses and toxins produced by microorganisms are all possible contaminants of food. However, microorganisms and their products can also be used to combat these pathogenic microbes. Probiotic bacteria, including those that produce bacteriocins, can kill and inhibit pathogens. Alternatively, purified bacteriocins such as nisin can be added directly to food products. Finally, bacteriophages, viruses that only infect bacteria, can be used to kill bacterial pathogens.

Thorough preparation of food, including proper cooking, eliminates most bacteria and viruses. However, toxins produced by contaminants may not be heat-labile, and some are not eliminated by cooking. Source of Contamination should be identified and have eradicate the problem and much more precautions should be taken to eliminate the contamination in foods.

FERMENTATION

Fermentation is one way microorganisms can change a food. Yeast, especially Saccharomyces cerevisiae, is used to leaven bread, brew beer and make wine. Certain bacteria, including lactic acid bacteria, are used to make yogurt, cheese, hot sauce, pickles, fermented sausages and dishes such as kimchi. A common effect of these fermentations is that the food product is less hospitable to other microorganisms, including pathogens and spoilage-causing microorganisms, thus extending the food's shelf-life.

Food fermentations are ancient technologies that harness microorganisms and their enzymes to improve the human diet. Fermented foods keep better, have enhanced flavours, textures and aromas, and may also possess certain health benefits, including superior digestibility. For vegetarians, fermented foods serve as palatable, protein-rich meat substitutes. Some cheese varieties also require molds to ripen and develop their characteristic flavours. Asian cuisines rely on a large repertoire of fermented foods. In particular, Aspergillus oryzae and A. sojae, sometimes called koji molds, are employed in many ways. Their hydrolytic enzymes suit them for growth on starch and other carbohydrate-rich substrates. In the koji process, fungal enzymes perform the same function as the malting enzymes used in the beer fermentations of western cultures.

The koji molds release amylases that break down rice starch, which in turn can be fermented to make rice wine. Fermented rice beverages have numerous local variations and names, depending on country and region. Rice wine is called shaoshing in parts of China, sake in Japan, takj or yakju in Korea, as well as by many other names across Asia. The koji molds are also effective in a variety of legume fermentations, of which miso and soy sauce are best known.

Miso is a mixture of soybeans and cereals usually used to flavour soups. Soy sauce is a flavourful, salty liquid sauce made from soybeans that have been fermented by koji molds, yeasts, as well as several halophilic bacteria. Other names for soy sauce include jiangyou (China), makjang and kanjang (Korea), toyo (Philippines) and siiu (Thailand).

Probiotics

Probiotics are living organisms that, when consumed, have beneficial health benefits outside their inherent nutritional effects. There is a growing body of evidence for the role of probiotics in gastrointestinal infections, irritable bowel syndrome and inflammatory bowel disease.

Lactobacillus species are used for the production of yogurt, cheese, sauerkraut, pickles, beer, wine, cider, kimchi, chocolate and other fermented foods, as well as animal feeds such as silage. In recent years, much interest has been shown in the use of lactobacilli as probiotic organisms and their potential for disease prevention in humans and animals.Bifidobacteria are considered as important probiotics, and are used in the food industry to relieve and treat many intestinal disorders.

Bifidobacteria exert a range of beneficial health effects, including the regulation of intestinal microbial homeostasis, the inhibition of pathogens and harmful bacteria that colonize and/or infect the gut mucosa, the modulation of local and systemic immune responses, the repression of procarcinogenic enzymatic activities within the microbiota, the production of vitamins, and the bioconversion of a number of dietary compounds into bioactive molecules.

MICROBIAL BIOPOLYMERS

A variety of biopolymers, such as polysaccharides, polyesters and polyamides, are naturally produced by microorganisms. Several microbially-produced polymers are used in the food industry.

Xanthan

Plant-pathogenic bacteria of the genus Xanthomonas are able to produce the acidic exopolysaccharide xanthan gum. Because of its physical properties, it is widely used as a viscosifer, thickener, emulsifier or stabilizer in the food industry. Xanthan consists of pentasaccharide repeat units composed of D-glucosyl, D-mannosyl, and D-glucuronyl acid residues in a molar ratio of 2:2:1 and variable proportions of O-acetyl and pyruvyl residues.

Alginate

Alginate is the main representative of a family of polysaccharides that neither show branching nor repeating blocks or unit patterns and this property distinguishes it from to other polymers like xanthan or dextran. Alginates can be used as thickening agents. Although listed here under the category 'Microbial polysaccharides', commercial alginates are currently only produced by extraction from brown seaweeds such as Laminaria hyperborea or L. japonica.

Cellulose

Cellulose is a simple polysaccharide, in that it consists only of one type of sugar (glucose), and the units are linearly arranged and linked together by ?-1,4 linkages only. The mechanism of biosynthesis is, however, rather complex, partly because in native celluloses, the chains are organized as highly ordered water-insoluble fibres. Currently, the key genes involved in cellulose biosynthesis and regulation are known in a number of bacteria, but many details of the biochemistry of its biosynthesis are still not clear. In spite of the enormous abundance of cellulose in plants, bacterial celluloses are being investigated for industrial exploitations.

Poly-γ-glutamic Acid

Poly-γ-glutamic acid (ã-PGA) produced by various strains of Bacillus has potential applications as a thickener in the food industry.

Levan

Levan, a homopolysaccharide composed of D-fructofuranosyl residues joined by 2,6 with multiple branches by 2,1 linkages, has great potential as a functional biopolymer in foods, feeds, cosmetics, and the pharmaceutical and chemical industries. Levan can be used as food or a feed additive with prebiotic and hypocholesterolemic effects.

Exopolysaccharides

Microorganisms synthesize a wide spectrum of multifunctional polysaccharides, including intracellular polysaccharides, structural polysaccharides and extracellular polysaccharides or exopolysaccharides (EPSs). EPSs generally consist of monosaccharides and some noncarbohydrate substituents (such as acetate, pyruvate, succinate, and phosphate). Owing to the wide diversity in composition, they have found multifarious applications in various food and pharmaceutical industries.

AGRICULTURAL BIOTECHNOLOGY AND FOOD SAFETY

When a consumer buys food s/he expects it to be safe and have the same quality characteristics that s/he is accustomed to, and familiar with. Consumers have the right to know that this is the case and if there is anything different about that product.

The Food and Agriculture Organization and the World Health Organization of the United Nations advocate the concept of 'substantial equivalence' as the most practical approach to address the safety evaluation of foods or food components derived through modern biotechnology. This approach states that 'if a new food or food component is found to be substantially equivalent to an existing food or food component, it can be treated in the same manner with respect to safety.'

So what kind of tests are used to determine if the food is as safe or safer than existing foodstuffs? Researchers must prepare comprehensive data to support the safety and wholesomeness of new crop varieties developed through biotechnology.

This process requires years of laboratory and field testing before a product can be brought to the market. To provide assurance that foods derived through biotechnology are as safe as those produced by traditional breeding programmes, the safety assessment strategies involve several key steps. These steps include molecular characterisation of the genetic modification, agronomic characterisation, nutritional assessment, toxicological assessment and safety assessment.

For example, typical questions that must be addressed are:

- Does the genetically modified food have a traditional counterpart that has a history of safe use?
- Has the concentration of any naturally occurring toxins or allergens in the food changed?
- Have the levels of key nutrients changed?
- Do new substances in the genetically modified food have a history of safe use?
- Has the food's digestibility been affected?
- Has the food been produced using accepted, established procedures?

Even after these and other questions about the biotechnology derived foods are answered, there are still more steps in the approval process before the crop can be commercialized.

In fact, genetically modified foods are the most studied food products ever produced:

- *Safety assessment*: Molecular characterisation – for new plant varieties produced through modern biotechnology, the source of the gene introduced into the plant is first identified. The transformation system used to insert the gene is defined, as well as the number of copies of inserted genes and the integrity and stability of the genetic insert.
- *Agronomic traits*: Usually the starting points for evaluating substantial equivalence. For example, in the case of potatoes, the traits commonly examined are yield, tuber size and distribution, dry matter content and disease resistance.
- *Nutritional assessment*: Involves key nutrients including fats, proteins, carbohydrates and essential vitamins and minerals.
- *Toxicology assessment*: Toxicants and anti-nutrients are compounds known to be naturally present in some crops that could have an impact on health if their levels increased. For example, solanine glycoalkaloids in potatoes or trypsin inhibitors in soybeans). The levels of anti-nutrients in crops produced through biotechnology are compared to conventionally produced varieties grown under comparable environmental and agronomic conditions.

- *Safety assessment*: When a crop produced through biotechnology is shown to be substantially equivalent to a conventional crop, the safety assessment focuses on the introduced trait and the expressed protein product.

The overall goal of these tests is to determine whether the plant is substantially equivalent to food derived from a conventional source that has a history of safe use. A substantial equivalence evaluation focuses on the product rather than the process used to develop the product. If the new product is substantially equivalent to the conventional food or feed, then the product derived through biotechnology is considered to be as safe as the conventional counterpart. If the food produced using biotechnology contains a new trait, which changes the levels of nutrients or antinutrients, such as a higher level of a vitamin or a lower level of an allergen, the assessment focuses on demonstrating the safety of the new trait.

ALLERGENS

One of the public's biggest concerns related to genetically modified foods is that an allergen could be accidentally introduced into a food product. The DNA inserted into a plant is safe and no case of allergy to DNA is known, the concern is whether the protein produced by the plant as a result of the inserted DNA could cause an unexpected allergic reaction.

Scientists know a lot about which foods trigger allergic reactions in adults and children. Ninety per cent of all food allergies are associated with only eight foods or food groups – shellfish, eggs, fish, milk, peanuts, soybeans, tree nuts, and wheat. These, and many other food allergens, are well characterized. To date, genes found to express any known allergenic protein have been specifically excluded from the research and development process.

Scientists have agreed that following these simple steps provides compelling assurance that no allergic effects will result from foods derived from biotechnology:

- Avoid transferring genes from foods known to be allergenic.
- Check the structure of any new proteins produced in foods derived from biotechnology against the structures of known allergens to assure that no allergenic structures exist in the new protein.
- Measure the stability of the new protein in stomach and intestinal fluids. Most allergens are stable to these conditions. Proteins which are unstable to these conditions are not likely to be allergens.
- Determine how much of the new protein will be present in the food consumed by humans. Most allergens are present in large amounts.

Allergenicity screening continues to be a very important part of safety testing before a crop can enter into the food market.

ANTIBIOTIC RESISTANCE

Some biotechnology crops contain genes for a trait called antibiotic resistance. Scientists use this trait as a marker to identify cells into which the desired gene has been successfully introduced. Concerns have been raised that these marker genes could move from biotechnology crops to microorganisms that normally reside in a person's gut and lead to an increase in antibiotic resistance.

There have been numerous scientific reviews and experimental studies of this issue and they have come to the following conclusions:

- The likelihood of antibiotic resistance genes moving from biotechnology crops to other organisms is extremely remote.
- In the unlikely event that an antibiotic resistance gene is transferred to another organism, the impact of this transfer would be negligible, as the markers used in genetically modified crops have limited or no clinical or veterinary use.

However, in response to public concerns, scientists have been advised to avoid using antibiotic resistance genes in biotechnology modified plants. Alternative marker strategies are being developed.

GENETICALLY MODIFIED FOOD CROPS

Genetic modification, otherwise referred to as recombinant DNA (rDNA) technology or gene-splicing, has proven to be a more precise, predictable and better understood method for the manipulation of genetic material than previously attained through conventional plant breeding. To date, agricultural applications of the technology have involved the insertion of genes for desirable agronomic traits (e.g. herbicide tolerance, insect resistance) into a variety of crop plants, and from a variety of biological sources.

Examples include soybeans modified with gene sequences from a *Streptomyces* species encoding enzymes that confer herbicide tolerance, and corn plants modified to express the insecticidal protein of an indigenous soil microorganism, *Bacillus thuringiensis* (Bt). A growing body of evidence suggests that the technology may be used to make enhancements to not only the agronomic properties, but the food, nutritional, industrial and medicinal attributes of genetically modified (GM) crops.

Regulatory supervision of rDNA technology and its products has been in place for a longer period of time in the United States than in most other parts of the world. The methods and approaches established to evaluate the safety of products developed using rDNA technology continue to evolve in response to the increasing availability of new scientific information.

As our understanding of the potential applications of the technology is broadened, the safety of products developed using rDNA technology and the potential effects of introduced gene sequences on human health or the environment will be more closely scrutinized. In fact, much of the knowledge acquired during the commercialization of the products of rDNA technology in agriculture is now finding application in evaluating the safety of products developed through more conventional means. The objective of this chapter is to provide the reader with an overview of the significant events leading up to the present, science-based, regulatory framework that exists for the safety evaluation of GM food crops within the United States. An attempt has been made to discuss concerns over the sufficiency of existing regulations, as well as to highlight recent initiatives taken by federal regulatory agencies to address them. Through better communication of how the regulatory process functions within the United States, it is anticipated that current and future applications of rDNA technology in agriculture will be met with a greater level of understanding and acceptance.

HISTORICAL PERSPECTIVE

rDNA technology was first developed in the 1970s. The initial response of the scientific community, including members of the National Academy of Science (NAS), to the prospects of rDNA technology, was to postpone any further research involving the technology until the potential risks to human health and the environment could be evaluated.

Researchers attending the International Conference on Recombinant DNA Molecules in 1975, otherwise known as the Asilomar Conference, tried to establish a scientific consensus on how best to self-regulate emerging applications of the technology. The conditions and restrictions that were proposed at this conference have formed the basis by which federal guidelines and policies for rDNA technology research were drafted within the United States.

National Institutes of Health (NIH)

The National Institutes of Health (NIH) was the first federal regulatory agency to publish their interests in evaluating the safety of rDNA technology in 1976, in the form of guidelines for the conduct of research. Because of the uncertainties that existed at the time, all research into the potential applications of rDNA technology was limited to the confines of federally funded labouratories under NIH control. After continued research, and a more careful assessment and monitoring of the risks, a set of less restrictive guidelines was published in 1978.

However, the environmental release of organisms developed using rDNA technology outside the confines of controlled labouratory conditions was prohibited unless otherwise approved by the NIH director. In the early 1980s, the NIH established an rDNA Advisory Committee (RAC) to review all data and experience gained with applications of the technology under its control. Based on recommendations of the RAC, a more relaxed set of research guidelines was published by the NIH in 1983.

The NIH approved the first environmental release of an organism developed using rDNA technology (ice-minus strain of *Pseudomonas*) in 1983.

In response, they were criticized for failing to prepare a statement or assessment of the environmental impact of their regulatory decision as required under the National Environmental Policy Act (NEPA). Once the legal controversy had subsided, all responsibility that the NIH had for regulating the environmental introduction of GM organisms was relinquished. Nevertheless, NIH guidelines continue to be referenced in assessing the safety of rDNA research performed within industry, federal and other state labouratories. However, it was unclear which federal regulatory agencies would be responsible for ensuring the safety of the products developed using rDNA technology.

FOOD SPOILAGE

Food is considered contaminated when unwanted microorganisms are present. Most of the time the contamination is natural, but sometimes it is artificial. Natural contamination occurs when microorganisms attach themselves to foods while the foods are in their growing stages. For instance, fruits are often contaminated with yeasts because yeasts ferment the carbohydrates in fruits. Artificial contamination occurs when food is handled or processed, such as when fecal bacteria enter food through improper handling procedures. Food spoilage is a disagreeable change or departure from the food's normal state. Such a change can be detected with the senses of smell, taste, touch, or vision. Changes occurring in food depend upon the composition of food and the microorganisms present in it and result from chemical reactions relating to the metabolic activities of microorganisms as they grow in the food.

TYPES OF SPOILAGE

Various physical, chemical, and biological factors play contributing roles in spoilage. For instance, microorganisms that break down fats grow in sweet butter (unsalted butter) and cause a type of spoilage called rancidity. Certain types of fungi and bacteria fall into this category. Species of the Gram-negative bacterial rod Pseudomonas are major causes of rancidity. The microorganisms break down the fats in butter to produce glycerol and acids, both of which are responsible for the smell and taste of rancid butter. Another example occurs in meat, which is primarily protein. Bacteria able to digest protein (proteolytic bacteria) break down the protein in meat and release odoriferous products such as putrescine and cadaverine. Chemical products such as these result from the incomplete utilization of the amino acids in the protein.

Food spoilage can also result in a sour taste. If milk is kept too long, for example, it will sour. In this case, bacteria that have survived pasteurization grow in the milk and produce acid from the carbohydrate lactose in it. The spoilage will occur more rapidly if the milk is held at room temperature than if refrigerated. The sour taste is due to the presence of lactic acid, acetic acid, butyric acid, and other food acids.

SOURCES OF MICROORGANISMS

The general sources of food spoilage microorganisms are the air, soil, sewage, and animal wastes. Microorganisms clinging to foods grown in the ground are potential spoilers of the food. Meats and fish products are contaminated by bacteria from the animal's internal organs, skin, and feet. Meat is rapidly contaminated when it is ground for hamburger or sausage because the bacteria normally present on the outside of the meat move into the chopped meat where there are many air pockets and a rich supply of moisture. Fish tissues are contaminated more readily than meat because they are of a looser consistency and are easily penetrated. Canned foods are sterilized before being placed on the grocery shelf, but if the sterilization has been unsuccessful, contamination or food spoilage may occur. Swollen cans usually contain gas produced by members of the genus Clostridium. Sour spoilage without gas is commonly due to members of the genus Bacillus.

This type of spoilage is called flat-sour spoilage. Lactobacilli are responsible for acid spoilage when they break down the carbohydrates in foods and produce detectable amounts of acid. Among the important criteria determining the type of spoilage are the nature of the food preserved, the length of time before it is consumed, and the handling methods needed to process the foods. Various criteria determine which preservation methods are used.

RISK OF GMOS AND GM FOODS TO HUMAN HEALTH AND THE ENVIRONMENT

Introduction of a transgene into a recipient organism is not a precisely controlled process, and can result in a variety of outcomes with regard to integration, expression and stability of the transgene in the host.

HISTORY OF RISK ASSESSMENT OF GMOS

When new foods are developed by traditional breeding methods, they are usually not subject to specific pre- or postmarket risk or safety assessment by national authorities or through international standards. This is in contrast to requirements introduced for GMOs and GM foods. The concept of risk assessment of GMOs was first discussed at the Asilomar Conference in 1975. The discovery of recombinant DNA had raised concerns among researchers regarding the potential creation of recombinant viruses whose escape would threaten public health.

Fourteen months after a voluntary moratorium on research involving recombinant DNA techniques, guidelines for the physical and biological containment of riskier experiments were drafted and agreed. These guiding principles were the basis of the USA guidelines for research in modern biotechnology developed in 1976 by the National Institutes of Health Recombinant DNA Advisory Committee. Other countries were soon to follow.

Early regulatory requirements were intended to prevent the accidental release of microorganisms from research facilities. In continuation of this, regulation for contained use and deliberate release of GMOs was developed, *e.g.* EU regulations in 1990. These guidelines elaborated a premarket human-health and environmental-safety assessment requirement for all GMOs and GM foods on the basis that they are novel and have no history of safe food or environmental use. Many countries have since established specific premarket regulatory systems requiring the rigorous assessment of GMOs and GM foods before their release into the environment and/or use in the food supply.

A summary of some national and international legislation is available on the OECD Internet site. While many national regulatory bodies base their safety assessment of GMOs and GM foods on shared concepts, differences in regulatory systems have led to disagreements and confusion in their deployment. While the terms 'safety assessment' and 'risk assessment' are often used interchangeably in some literature, these are two clearly different, but interlinked processes. To provide international consistency in risk analysis of GMOs and GM foods which incorporates risk assessment, management and communication components, a number of international regulatory and standard-setting bodies have introduced uniform standards.

These include standards for human-health and environmental-safety assessment of GMOs and GM foods, and notification of their movement across national borders. The objective of uniform global standards for risk assessment would be challenging as countries are bound to reach different decisions on the scope of the assessment, particularly the resolution of whether or not to include social or economic aspects. International regulatory systems covering GM food safety and environmental safety came into force in 2003. The concept that allows for the comparison of a final product with one having an acceptable standard of safety is an important element of a GM food safety assessment. This principle was elaborated by FAO, WHO and OECD in the early 1990s and referred to as 'substantial equivalence'.

The principle suggests that GM foods can be considered as safe as conventional foods when key toxicological and nutritional components of the GM food are comparable to the conventional food and when the genetic modification itself is considered safe.

However, the concept has been criticized by some researchers. At a *Joint FAO/WHO consultation on foods derived from biotechnology* held in 2000, it was acknowledged that the concept of substantial equivalence contributes to a robust safety assessment, but it was also clarified that the concept should represent the starting point used to structure the safety assessment of a GM food relative to its conventional counterpart.

The consultation concluded that a consideration of compositional changes should not be the sole basis for determining safety, and that safety can only be determined when the results of all aspects under comparison are integrated.

This study does not cover aspects of occupational health which are often addressed in regulations dealing with the safety of work with GMOs in contained areas. It should also be noted that the adventitious presence of non-approved products of modern biotechnology among the approved is not within the scope of this study.

ASSESSMENT OF THE IMPACT OF GM FOODS ON HUMAN HEALTH

Principles for the Safety Assessment of GM Foods

The Codex Alimentarius Commission adopted the following texts in July 2003: *Principles for the risk analysis of foods derived from modern biotechnology*; *Guideline for the conduct of food safety assessment of foods derived from recombinant-DNA plants*; and *Guideline for the conduct of food safety assessment of foods produced using recombinant-DNA microorganisms*. The last two texts are based on the *Principles* and describe methodologies for conducting safety assessments for foods derived from recombinant-DNA plants and microorganisms, respectively. The premise of the *Principles* dictates a premarket assessment, performed on a case-by-case basis and including an evaluation of both direct effects and unintended effects.

The Codex safety assessment principles for GM foods require investigation of:

- Direct health effects;
- Tendency to provoke allergic reactions;
- Specific components thought to have nutritional or toxic properties;
- Stability of the inserted gene;
- Nutritional effects associated with the specific genetic modification; and
- Any unintended effects which could result from the gene insertion.

Codex principles do not have a binding effect on national legislation, but are referred to specifically in the *Agreement on the Application of Sanitary and Phytosanitary Measures* of the World Trade Organization and are often used as a reference in the case of trade disputes.

The 2003 *Expert consultation on the safety assessment of foods derived from GM animals, including fish* formed the opinion that to further develop the risk-assessment process with current scientific knowledge, integrated toxicological and nutritional evaluations should be conducted in order to identify food-safety issues that may need further investigation. Both evaluations combine data from the hazard identification and characterization, and food intake assessment steps.

It should be noted that such newly suggested further developments of the risk-assessment process have not yet been considered by Codex, and that the international principles and guidelines for risk analysis and safety assessment of foods derived from biotechnology are as accepted by Codex in 2003.

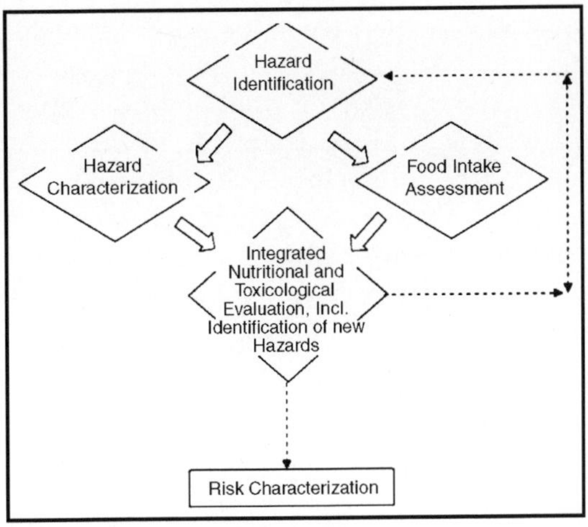

Fig. Schematic overview of a Suggested Further Development of the Risk Assessment Process

Potential Direct Effects on Human Health

The potential direct health effects of GM foods are generally comparable to the known risks associated with conventional foods, and include, for example, the potential for allergenicity and toxicity of components present, and the nutritional quality and microbiological safety of the food. Many of these issues have not traditionally been specifically assessed for conventional food; but in one area—toxicity of food components—there is ample experience related to the use of animal experiments to test potential toxicity of targeted chemical components.

However, the intrinsic difficulty in testing whole foods, as opposed to specific components, in animal feeding experiments have resulted in the development of alternative approaches for the safety assessment of GM foods. The safety assessment of GM food follows a stepwise process aided by a series of structured questions.

Factors taken into account in the safety assessment include:

- Identity of gene of interest, including sequence analysis of flanking regions and copy number;
- Source of gene of interest;
- Composition of GMO;
- Protein expression product of the novel DNA;
- Potential toxicity;
- Potential allergenicity; and
- Possible secondary effects from gene expression or the disruption of the host DNA or metabolic pathways, including composition of critical macronutrients, micronutrients, antinutrients, endogenous toxicants, allergens and physiologically active substances.

A series of FAO/WHO expert consultations held in 2000, 2001 and 2003 recognized that animal studies can be of help but that there are practical difficulties in obtaining meaningful information from conventional toxicology testing, especially with whole-food studies in laboratory animals. The consultations also noted that very little is known about the potential long-term effects of any foods. At present, there is no conclusive information available on the possible health effects of modifications which would significantly change the nutritional characteristics of any food, such as nutritionally enhanced foods.

Potential Unintended Effects of GM Foods on Human Health

Unintended effects, such as elevated levels of antinutritional or toxic constituents in food, have on occasion been characterized in conventional breeding methods, *e.g.* glycoalkaloid levels in potatoes. Organisms derived from conventional breeding methods, including tissue cultures, may have a somewhat enhanced possibility for genetic instabilities, such as the activity of mobile elements and gene-silencing effects.

These effects could increase the probability of unintended pleiotropic effects, *e.g.* increased or decreased expression of constituents or possibly modifications in expressed proteins, as well as epistasis. It has been argued that random insertion of genes in GMOs may cause genetic and phenotypic instabilities but, as yet, no clear scientific evidence for such effects is available.

A better understanding of the impact of natural transposable elements on the eukaryotic genome may shed some light on the random insertion of sequences. Gene expression in conventional and GM crops is subject to environmental influences. Environmental conditions such as drought or heat can stimulate some genes; turning the expression up or down.

The assessment of potential synergistic effects is necessary in the risk assessment of organisms derived from gene stacking, *i.e.* breeding of GMOs containing genetic constructs with multiple traits. Internationally agreed procedures for the assessment of such organisms are desirable. Unintended effects can be classified as insertional effects, *i.e.* related to the position of insertion of the gene of interest, or as secondary effects, associated with the interaction between the expressed products of the introduced gene and endogenous proteins and metabolites.

There is common agreement that targeted approaches, *i.e.* the measuring of single compounds, is very useful and adequate to detect such effects, as has been done with conventionally bred products. To enhance and improve the identification and analyses of these unintended effects, profiling methods have been suggested. This untargeted approach allows detection of unintended effects at the mRNA, protein and metabolite level. It still remains to be seen which of these techniques would be useful for routine risk-assessment purposes. Unintended effects were specifically addressed by the FAO/WHO Expert consultation on the safety aspects of genetically modified foods of plant origin and the Codex Principles for the risk analysis of foods derived from modern biotechnology.

These consultations noted that there is a need to establish the consequences of natural baseline variations, the effects of growing conditions and environmental influences, and the ways to interpret safety-relevant data from profiling techniques. Adequate methods for the assessment of potential, unintended effects need to be evaluated for specific GMOs case by case, where the assessment already aims to consider unintended toxic and antinutritional factors through analysis of proximal constituents and GM characteristics. As profiling methods are not in use in routine risk assessment, the second step in the comparative safety assessment has been suggested as a measure for identifying and characterizing any unintended effects that may be associated with complex foods.

Potential Human-Health Effects from Horizontal Gene Transfer

Natural genetic transformation has been found to occur in different environments, *e.g.* in food. In addition, it has been shown that ingested DNA from food is not completely degraded by digestion, and that small fragments of DNA from GM foods can be found in different parts of the gastrointestinal tract. As the consequences of horizontal gene transfer may be significant in some human-health conditions, the potential for HGT needs to be part of the risk assessment of GM food. FAO/WHO consultations have also discussed the potential risks of gene transfer from GM foods to mammalian cells or gut bacteria.

These panels have suggested that it may be prudent in a food-safety assessment to assume that DNA fragments survive in the human gastrointestinal tract and can be absorbed by either the gut microflora or somatic cells lining the intestinal tract. It was agreed that the assessment needs to take into account a number of factors including, but not limited to, the specific characteristics encoded by the DNA sequences, the characteristics of the receiving organism, and the selective conditions of the local environment of the receiving organisms. Some scientists have pointed to the present methodological limitations of a comprehensive scientific evaluation of this problem. Discussion also addresses the consequences of a rare probability of a transfer event against the high numbers of bacteria and genes available for transfer.

The DNA construct used to change the genetic composition of a recipient organism should be considered within an assessment, especially if the gene or its promoter has been derived from a viral source. Sequences unrelated to the target gene could be introduced as part of the construct. Inadvertent introduction of such sequences into the germ-line of a GM animal not only has the potential for creating unintended genetic damage, but can also contribute by recombination to the generation of novel infectious viruses. A well-known example is the generation of a replication-competent murine leukaemia virus during the development of a vector containing a globin gene.

The horizontal transfer of recombinant genetic material to microorganisms has demonstrated an enhanced stability of DNA under certain conditions. Natural transformation of DNA to bacteria involves the active uptake of extracellular DNA by bacteria in a status of competence or in rare, illegitimate recombination events.

The probability of such an event occurring appears to be extremely low, and very much related to the genes, constructs and organisms in question.

The FAO/WHO expert panels concluded that horizontal gene transfer is a rare event that cannot be completely discounted, and that the consequences of such transfer should be considered in a safety assessment.

The panels encouraged the use of recombinant DNA without antibiotic-resistance genes or any other sequences which could stimulate transfer. The panels also discouraged the use of any unnecessary DNA sequences, including marker genes in the genetic construct. The safety assessment of a genetic construct should also examine the included marker genes.

Commonly used marker genes code for antibiotic resistance. Risk assessment of these selectable genes should focus on gene transfer to microorganisms residing in the gastrointestinal tract of humans or animals. As the potential of this gene transfer cannot be completely ruled out, the safety assessment should also consider information on the role of the antibiotic in human and veterinary medical uses.

Potential Immune Responses and Allergenicity Iinduced by GM Foods

Food allergies or hypersensitivities are adverse reactions to foods triggered by the immune system. Within the different types of reactions involved, non-immunological intolerances to food and reactions involving components of the immune system need to be differentiated. The former may invoke reactions such as bloating or other unpleasant reactions, but are thought not to involve the immune system and called 'food intolerances'.

Allergic reactions to traditional foods are well known. The major food allergens are proteins in and derived from eggs, fish, milk, peanuts, shellfish, including crustaceans and molluscs, soy, tree nuts and wheat. Whereas the groups of main allergens are well known and advanced testing methods have been elaborated, traditionally developed foods are not generally tested for allergens before market introduction. The application of modern biotechnology to crops has the potential to make food less safe if the newly added protein proves to cause an allergic reaction once in the food supply. A well-known case is the transfer of a gene encoding a known allergen, the 2S-Albumin gene from the Brazil nut, to a previously safe soybean variety.

When the allergenic properties of the transgenic soybean were tested, sera from patients allergic to Brazil nuts cross-reacted with the transgenic soybean.

For this reason, a commercial product was never pursued. On the other hand, the introduction of an entirely new protein that has not been previously found in the food chain represents a different case. In the first case, guidelines for assessing foods with known allergens are clear. The second case is more difficult to assess because there is no definitive test to determine the potential allergenicity of a novel protein. Instead, several risk factors provide a rough guide as to the likelihood of allergenicity.

Risk-assessment protocols for food allergy examine four elements:

1. Allergenicity assessment
2. Dose response assessment
3. Exposure assessment and
4. Susceptible subpopulations.

Elements of an allergenicity assessment include a comparison of the sequence of the transferred gene with sequence motifs of allergenic proteins from databanks, an evaluation of the stability of the newly expressed proteins against digestion, and animal and immune tests, as appropriate. Absence of sequence similarity with allergenic protein epitopes, and low stability under acidic or proteolytic conditions, do not preclude the presence of a potential allergen.

There are proven incidents which have contradicted the general rules, *e.g.* where small modifications in a protein sequence determine allergenicity. Allergenicity prediction using protein-sequence motifs identified from a new allergen database has been proposed as a new and superior strategy for identifying potential allergens. Some experts consider that the use of sera from polysensitized patients is important for the testing of allergenicity.

Areas of improvement of risk assessment of allergens include mechanistic studies of animal models and genomic techniques. FAO/WHO expert panels have established protocols for evaluating the allergenicity of GM foods on the basis of the weight of evidence. The strategy adopted is applicable to foods containing a gene derived from either a source known to be allergenic or a source not known to be allergenic. The panels have, however, discouraged the transfer of genes from known allergenic foods unless it can be demonstrated that the protein product of the transferred gene is not allergenic.

These principles have been applied by many regulatory agencies assessing the safety of GM foods and have provided the basis for Codex guidelines for the safety assessment of foods derived from biotechnology. The cellular basis of immune responses is not completely understood, and a better understanding of the interaction of the immune system and foods in general is required in order to decipher whether specific GM foods may have impacts on the immune system apart from allergenicity. The impact of cell-mediated reactions on hypersensitivity reactions elicited by foods is a matter of current research.

Safety Aspects of Food Derived from GM Animals

Genetically modified animals have mainly been produced for biomedical research purposes. To date, no GM food animals have been introduced onto international markets. But GM food animals such as fish can be expected in the near future. In principle, the assessment of food and feed safety for GM animals follows the general principles of the assessment of GMOs. However, the specificities of the introduction of transgenes into animals, often using viral constructs for introduction into the germ-line, need distinct consideration. A 2003 report of the Pew Initiative on Food and Biotechnology reviewed techniques for the production, uses and welfare of GM animals, as well as safety aspects.

The risk assessment of foods derived from GM animals needs to be undertaken, as for other GM foods, on a case-by-case basis. This includes an assessment of potential recombination of viral vectors used for transformation with wild-type viruses, especially in poultry, where potential incomplete digestion could lead to intestinal uptake of orally administered proteins, and an assessment of peptide expression that may have hormonal activity. The FAO/WHO expert consultation on the *Safety assessment of foods derived from GM animals, including fish* held in 2003 addressed the key issues for food safety and evaluated the extent of scientific knowledge with regard to hazard identification and characterization unique to transgenic animals.

Phenotypic Analysis

Because of their size, and limitations in the generation process, there will in general be few initial founders for screening of GM animals, meaning that information on the variation range between animals with the same genetic modification will be rather limited. This will make interpretation of differences difficult.

Furthermore, a selection of the edible tissues and products to be analysed has to be made for the different animal species. In specific cases, phenotypic analysis may also be advisable after processing or, for fish, during the various stages of spoilage. For example, adverse biogenic amines can be formed during spoilage in salmon, tuna, herring and other fish species. Similarly, formaldehyde may be produced in spoiled shrimp, cod, hake and many other species.

Compositional Analysis

Background data on the natural variation for individual constituents in different tissues need to be generated. Data in existing databases must be evaluated for their quality and value for use in comparative compositional analysis.

Safety Aspects of Foods Derived from or Produced with GMMs

The production of food additives or processing aids using GMMs, where the microorganism is not a part of the food, has become an important and generally well-accepted technology, with a significant number of such products on the market. Experience with the purification of proteins in the biomedical field suggests that well-standardized purification protocols are of central importance for the safety of these products.

Where the GMMs are a part of the food matrix, certain criteria were established in 2001 by a *Joint FAO/WHO expert consultation on foods derived from biotechnology* for assessing the risks that may be associated with the preparation of such foods. These include the genetic constructs used in the GM microorganisms, the pathogenic potential of GMM, and the detrimental effects of a potential gene transfer. For GMMs used in foods, the ensuing risk assessment ought to focus on the effects of a possible interaction between the GMM and endogenous intestinal microflora and the potential immune-stimulatory or immunomodulatory effects of the microorganisms in the event the gastrointestinal tract is colonized.

Small regulatory elements derived from viral DNA are commonly used to drive the expression of transgenes in GMOs. Viral-DNA constructs are sometimes used as transgenes to establish resistance against viral pests, as they express viral proteins that confer viral resistance on plants.

Some scientists suggest that the potential interaction of GM viral constructs with related wild-type viruses needs to be part of the risk assessment, to evaluate the potential of new viral pest strains evolving through mechanisms of recombination. Insertion of viral vectors into functionally important genes of recipient patients in the field of biomedicine has been reported, and whereas such vectors are not commonly used in food production, this evidence indicates the limited understanding of mechanisms directing insertion of genetic constructs.

Safety Aspects of Foods Derived from Biopharming

The potential to produce human proteins in animals has resulted in great interest in new possibilities for human health, but also in efforts to establish appropriate risk-assessment methods. The biosafety aspects of molecular 'farming' can be divided into two major groups: the potential spread of transgenes; and the potential negative effects of the expressed protein on the environment and the consumer. Practices and guidelines ensuring effective separation of 'biopharming' are being investigated. Experts agree that the risk assessment should ensure that proteins designed to produce pharmaceutical products, *e.g.* in the animal's milk, cannot find their way to other parts of the animal's body, possibly causing adverse effects.

Potential Effect of GMOs on Human Health Mediated Through Environmental Impact

Work on environmental-health indicators suggests that various agricultural practices have direct and indirect effects on human health and development. Hazards can take many forms—wholly natural in origin, or derived from human activities and interventions. The need to assess indirect effects of the use of GMOs in food production has been emphasized by many countries.

Potential environmental-health hazards from the release of GMOs into the environment have been discussed in a report by WHO and the Italian Environment Protection Agency, in which health effects have been analysed "as an integrating index of ecological and social sustainability". For example, the production of chemicals or enzymes from contained GM microorganisms have contributed significantly to decreases in the amount of energy use, toxic and solid wastes in the environment, thereby significantly enhancing human health and development.

A further example of beneficial human/environmental outcomes of introducing GM crops is the reduction in the use, environmental contamination and human exposure to pesticides demonstrated in some areas. This has occurred especially through the use of pesticide-resistant *Bt* cotton, which has been shown to decrease pesticide poisoning in farm workers. Outcrossing of GM plants with conventional crops or wild relatives, as well as the contamination of conventional crops with GM material, can have an indirect effect on food safety and food security by contamination of genetic resources.

Although initial concerns about introgression of transgenic DNA into traditional landraces of maize in Mexico arose as a result of the findings of transgenic DNA in such landraces in 2000 recently published results from samples taken during a broad, systematic survey in 2003 and 2004 in the same region shows no transgenes in these landraces. Still, the potential for introgression remains a possibility and risk-mitigation measures are being considered.

Both outcrossing and contamination characteristics are dependent on the pollination and distribution characteristics of pollen and seeds of the specific plant. In the USA, GM 'Starlink' maize was not approved for food use, but unintentionally started to appear in maize food products. This example demonstrated the problem of contamination and highlighted the potential for unintended impacts on human health and safety.

In the case of Starlink maize, full segregation of GM varieties not intended for food use and other varieties of the same crop species could not be achieved. Improved molecular methods for containment of the transgenes as well as farm management measures are under discussion, *e.g.* isolation distances, buffer zones, pollen barriers, control of volunteer plants, crop rotation and planting arrangements for different flowering periods, and monitoring during cultivation, harvest, storage, transport and processing. The likelihood of GM animals entering and persisting in the environment will vary among taxa, production systems, modified traits, and receiving environments.

The spread and persistence of GM fish and shellfish—or their transgenes—in the environment could be an indirect route of entry of GM animal products into the human food supply. This is because escaped individuals or their descendents could subsequently be captured in fishing for those species. Similar mechanisms might apply for poultry such as ducks and quail that are subject to sport or subsistence harvest. Live transport and sale of GM fish and poultry pose another route for the escape of GM animals and their entry into the environment.

FOOD PRESERVATION

Food preservation methods are intended to keep microorganisms out of foods, remove microorganisms from contaminated foods, and hinder the growth and activity of microorganisms already in foods. To keep microorganisms out of food, contamination is minimized during the entire food preparation process by sterilizing equipment, sanitizing it, and sealing products in wrapping materials.

Microorganisms may be removed from liquid foods by filtering and sedimenting them or by washing and trimming them. Washing is particularly valuable for vegetables and fruits, and trimming is useful for meats and poultry products.

HEAT

When heat is used to preserve foods, the number of microorganisms present, the microbial load, is an important consideration. Various types of microorganisms must also be considered because different levels of resistance exist. For example, bacterial spores are much more difficult to kill than vegetative bacilli. In addition, increasing acidity enhances the killing process in food preservation.

Three basic heat treatments are used in food preservation: pasteurization, in which foods are treated at about 62°C for 30 minutes or 72°C for 15 to 17 seconds; hot filling, in which liquid foods and juices are boiled before being placed into containers; and steam treatment under pressure, such as used in the canning method. Each food preserved must be studied to determine how long it takes to kill the most resistant organisms present. The heat resistance of microorganisms is usually expressed as the thermal death time, the time necessary at a certain temperature to kill a stated number of particular microorganisms under specified conditions. In the canning process, the product is washed to remove soil. It is then blanched by a short period of exposure to hot water to deactivate enzymes in the food. Diseased parts in the food are removed, and the food is placed into cans by a filling machine. Sealed cans are then placed into a sterilizing machine called a retort, and the food is processed for a designated time and temperature.

LOW TEMPERATURES

Low temperatures are used to preserve food by lowering microbial activity through the reduction of microbial enzymes. However, psychrophilic bacteria are known to grow even at cold refrigerator temperatures. These bacteria include members of the genera Pseudomonas, Alcaligenes, Micrococcus, and Flavobacterium. Fungi also grow at refrigeration temperatures. Slow freezing and quick freezing are used for long-term preservation. Freezing reduces the number of microorganisms in foods but does not kill them all. In microorganisms, cell proteins undergo denaturation due to increasing concentrations of solutes in the unfrozen water in foods, and damage is caused by ice crystals.

CHEMICALS

Several kinds of chemicals can be used for food preservation, including propionic acid, sorbic acid, benzoic acid, and sulfur dioxide. These acids are acceptable because they can be metabolized by the human body. Some antibiotics can also be used, depending upon local laws and ordinances. Tetracycline, for example, is often used to preserve meats. Storage and cooking normally eliminates the last remnants of antibiotic. In many foods, the natural acids act as preservatives. In sauerkraut, for example, lactic acid and acetic acid prevent contamination, while in fermented milks (yogurt, sour cream), acids perform the same function. For centuries, foods were prepared in this manner as a way of preventing microbial spoilage.

DRYING

Drying is used to preserve food by placing foods in the sun and permitting the water to evaporate. Belt, tunnel, and cabinet dryers are used in industry for such things as instant coffee and cocoa. Freeze-drying, a process called lyophilization, is also valuable for producing a product free of moisture and very light.

RADIATIONS

Ultraviolet radiation is valuable for reducing surface contamination on several foods. This short-wavelength light has been used in the cold storage units of meat processing plants. Ionizing radiations such as gamma rays can be used to preserve certain types of vegetables, fruits, and spices.

WHY DO WE NEED TO PRESERVE FOODS

Now that you have understood the definition of food preservation, did you think about why we should preserve foods? Can you think of a few reasons for preserving foods? Let us find an answer to this question by taking an example of one food item. Let us take the example of a fruit, say mango. There are many ways by which mangoes can be preserved. These are: juice, murraba, squash, aam papad, pulp, chutney, pickle, raw mango powder.

Mango is a summer fruit and grows in large quantities during the months of April to August. Different varieties of mango are grown in different parts of our country. Usually all the quantity grown in a region cannot be consumed by the people staying there as there is always an excess. What does the farmer do with this excess quantity? He makes arrangements to transport the excess quantity to regions where either mango is not grown or where that particular variety of mango is not available. If he does not do this, the excess produce will rot and go waste. The farmer will then lose money. There is still some quantity which is left after the fresh fruit is consumed by the people. It is this quantity which has to be preserved for consumption during the months when mango is not available.

Preservation of foods is done during the months when food is available in large quantity and therefore at low cost:

- One of the important reasons for preserving foods is to take care of the excess produce. There are many other reasons for preserving foods. Let us learn about these.
- The second reason for preserving foods is that they add variety to our meals. Have you ever got tired of eating the same vegetables which are in season? Is it not nice to eat peas when they are either very expensive in the market or are not available ? Eating cauliflower in pulav or cauliflower vegetable during the summer months adds to the interest in meals. In the same way, eating some chatni, papad or pickle along with the meals adds to the variety. Preserving foods when they are in season makes this possible.
- Reaches areas where the food item is not grown. In some areas of Rajasthan which are desert areas and in Himalayan regions that are covered with snow most of the time, very few foods can be grown. Availability of some preserved foods can add to the variety and nutritive value of meals. For example inclusion of dehydrated peas, green leafy vegetables, canned fruits etc., in the meals is a good idea in such areas.
- Makes transportation and storage of foods easier. Preservation of foods usually reduces bulk. This makes their transportation and

storage easier since it requires less space. For example, if you dry green leafy vegetables such as mint, methi, corriander, etc., their weight and volume reduces, thus making their storage easy.

WHY DOES FOOD GET SPOILT

The definition of food preservation states that preservation is keeping food in such a state that they do not get spoilt for a long period. Before we look at the reasons of food spoilage, let us understand, when is a food spoilt. When you keep bread outside the refrigerator for few days, a spongy growth is seen on it, which may be white, green on black in colour.

The bread thus gets spoilt due to growth of mould and becomes unfit for consumption. Likewise, if cooked dal or vegetables is left outside for sometime, it develops a bad smell and bubbles due to fermentation.

The dal and vegetables are thus spoilt and cannot be eaten. Can you now say when is a food spoilt? Food is said to be spoilt if there is rotting, *i.e.*, bad smell, fermentation, *i.e.*, bubbles/gas in the food or mould *i.e.*, spongy growth on the foodstuff. Formation of soft spots or soft brown spots on fruits and vegetables is also food spoilage. Why do foods get spoilt? If you know the reasons of food spoilage, you can remove these conditions while preserving food items. Foods get spoilt mainly due to the presence of micro-organisms, enzymes (present in foods), insects, worms, and rats.

Presence of Micro-organisms

Micro means small. Micro-organisms are very small organisms which cannot be easily seen. Micro-organisms spoil food items when the condition for their growth are appropriate.

What are these appropriate conditions? Like all living beings micro-organisms require air, moisture, right temperature and food to grow and multiply.

The situations which provide appropriate conditions for growth of micro-organisms, can be listed as:

- Food having high moisture content.
- Air around the food containing micro-organisms.
- Foods kept for a long time at room temperature.
- Skin of fruits and vegetables getting damaged, thus exposing the food to micro-organisms.
- Foods with low salt, sugar or acid content.

Presence of Enzymes

Enzymes are chemical susbtances found in all plants and animals. Are enzymes harmful to foods? No, enzymes help in ripening of fruits and vegetables. A raw green mango after a few days becomes sweet in taste and yellow in colour due to the enzymes action. What will happen if you keep this yellow, ripe mango for a few more days?

It will become soft, develop black spots and will start smelling bad. This is due to continued action of enzymes. No one likes to eat such as over ripe, spoilt mango. You know that even when the skin of fruits is not cut or damaged, it gets spoilt. This is due to enzyme action.

PRINCIPLES OF FOOD PRESERVATION

After learning about the causes of food spoilage, it should not be very difficult to list the principles of food preservation. Remember, a good method of food preservation is one that slows down or prevents altogether the action of the agents of spoilage.

Also, during the process of food preservation, the food should not be damaged. In order to achieve this, certain basic methods were applied using the knowledge gained form observation of the effects of natural conditions on different types of foods. For example in earlier days, in very cold weather condition, ice was used to preserve foods. Thus, very low temperature became an efficient method for preventing food spoilage.

Let us now list the principles of food preservation:

- *Removal of Micro-organisms or Inactivating Them*: This is done by removing air, water (moisture), lowering or increasing temperature, increasing the concentration of salt or sugar or acid in foods. If you want to preserve green leafy vegetables, you have to remove the water from the leave so that micro-organisms cannot survive. You do this by drying the green leaves till all the moisture evapourates.
- *Inactivating Enzymes*: Enzymes found in foods can be inactivated by changing their conditions such as temperature and moisture, when you preserve peas, one of the methods of preservations is to put them for a few minutes in boiling water. This method inactivates enzymes and thus, in preserving the food.
- *Removal of Insects, Worms and Rats*: By storing foods in dry, air tight containers the insects, worms or rats are prevented from destroying it.

CHEMICALS IN FOOD PROCESSING

The same over commitment to potentially risky methods is found in other phases of food production. Complex problems of storage, cleansing, handling, refining, cooking, mixing, heating and packaging are often solved by the use of chemical additives. A half century ago the yellowish tint in freshly milled flour was removed slowly by aging; today it is removed rapidly by oxidizing agents.

The miller no longer has to store flour and wait for it to mature. Emulsifiers are employed, partly to "improve" the texture of a product, partly to speed up processing. Their "functional advantages" in ice cream, for example, are that the "mix can be whipped faster, the product from the freezer is dryer and the mix has less tendency to 'churn' in the freezer."

Chemical antioxidants are used to help prevent oils and fats from becoming rancid, a problem that becomes especially pronounced when natural antioxidants are removed during large-scale processing. Chemical preservatives are added to food to permit longer storage in wholesale depots and to extend the "shelf life" of a product in retail outlets. In time chemicals are turned from mere adjuncts of food production into "technological necessities." Their use may result in new machines and facilities, the abandonment of old processing methods and a broad reorientation of technology to meet the new chemical requirements.

The fact that a chemical is technologically useful does not mean that it is biologically desirable. The two phrases are occasionally used interchangeably, as though they were synonymous. Some of the chemicals most widely used in the production of food have been found to be harmful to animals in laboratory tests. Nitrogen trichloride, or agene, was employed for decades as a bleaching and maturing agent in flour before it was discovered, in 1946, that the compound produces "running fits," or hysteria, in dogs.

The use of agene in the bread industry has since been discontinued. Although there is no evidence that agenized flour causes nerve disorders in man, no one can say with certainty that cereal products treated with agene are completely harmless to consumers, especially if eaten day after day for twenty years or more.

The toxicant in agenized flour "produces nervous disturbances in monkeys, but not epilepsy in man," observed the late Anton Carlson, one of the most distinguished physiologists of his day. "That is not sufficient for me, because there are many other types of nervous disturbances that may be aggravated or introduced by this chemical."

Another group of chemicals, notably certain types of polyoxyethylene derivatives, were widely used in the United States as emulsifiers before they were banned as potential hazards to consumers. Former F.D.A. Commissioner Charles W. Crawford is reported to have described the agents as "good paint removers." According to data furnished to the Delaney Committee, rats that had been fed two commercial preparations of polyoxyethylene-lb/> derived emulsifiers showed blood in their feces and developed a variety of kidney, abdominal, cardiac and liver abnormalities.

These preparations were used primarily as bread softeners but also found their way into other foods. The polyoxyethylene derivatives constitute a highly suspect family of emulsifying agents, yet some of them are still added to cake mixes, cake icing, frozen desserts, ice cream, candies, dill pickles, fruit and vegetable coatings, vitamin preparations and many food-packaging materials. Similar examples can be culled from other groups of chemical aids. In general, the toxic effects of many commonly used compounds are difficult to determine, but individually and in groups they have aroused suspicion and concern. "The fortification of oils and fats against oxidation and rancidity is another field in which there are serious questions of possible toxicity," warns F. J. Schlink, technical director of Consumers' Research and editor of *Consumer Bulletin*. "The natural antioxidants which delay rancidity are lost during the factory processing of refined oils and fats; then the attempt is made to restore the antioxidation qualities by addition of fat-stabilizing substances. Most of the materials that have antioxidation properties are known to be toxic to some degree."

The term "technological aid" has been used loosely to include food additives that could be dispensed with entirely if a manufacturer improved his processing methods and the quality of his product. The question of whether a chemical, instead of being legitimately employed to furnish distant consumers with a perishable product, was being used rather as a device for masking inferior ingredients and unsanitary methods arose in sharp form early in the century, when sodium benzoate was employed extensively as a preservative in catsup. Harvey Wiley, the first administrator of the federal food and drug law, regarded sodium benzoate as a hazard to consumers.

Supported by data from feeding experiments performed with human beings, Wiley claimed that the preservative causes injury to the digestive system. He was convinced that catsup need not deteriorate readily if clean facilities and proper ingredients are used in preparing it. Challenged on this point by Charles F. Loudon, a canner in Terre Haute, Indiana, Wiley appointed a bacteriologist, Arvill W. Bitting, to work at the Loudon factory and prepare a formula for a stable catsup preparation without artificial preservatives. Bitting was successful.

Moreover, after inspecting numerous canning factories and examining many brands of catsup, Bitting concluded that manufacturers who needed sodium benzoate often used inferior products and were careless about sanitation. "The whole spirit and tradition of the pure-food law is against the use of preservatives and substitutes," declared Mrs. Harvey Wiley more than forty years after the Loudon episode. "The intent of the law is to encourage the manufacture of high-grade ingredients, not to try to hide inferiority and cheapness by the use of chemicals and preservatives." If this is so, the spirit of the law has been ignored for decades.

The most up-to-date surveys of chemical additives in foods show that preservatives are added to cheese, margarine, cereal products, jams, jelly and many other processed foods. Uncooked poultry is dipped in solutions of antibiotics. Aside from the damage to public health that may arise from the use of questionable chemicals as preservatives, the longer "shelf life" acquired by some of these foods may cause a loss of valuable nutrients. Many chemical additives in foods perform no technological or nutritional function whatever. They cannot be regarded as materially useful by any stretch of the imagination. These chemicals enhance the colour of a food or make certain products feel soft and newly prepared; in some instances they have been used to replace costly but valuable nutrients with inferior ones. In the least objectionable cases, an additive will deceive the consumer without impairing his health.

The application of an innocuous vegetable dye to some foods often leads the consumer to believe that he is acquiring a better, more wholesome, or tastier product than is actually the case. In the most objectionable cases, a chemical is toxic to a greater or lesser degree. This type of additive not only serves to change the appearance or conceal the nutritive deficiency of a food; it exposes the consumer to a certain amount of damage.

Nitrates and nitrites, for example, are used to impart a pink colour to certain brands of processed meat. Ostensibly, this is their sole function. The addition of these chemicals to meat, especially frankfurters and hamburger meat, would be objectionable if only because of the deception which makes it difficult for the housewife to distinguish between high-quality and low-quality products. But this is not the only deception perpetrated on the consumer. Used together with salt, nitrite compounds definitely extend the "shelf life" of processed meat products.

When nitrites are ingested, they react with hemoglobin in the blood to form methemoglobin and, like carbon monoxide, reduce the hemoglobin's capacity to carry oxygen. An individual who consumes three to four ounces of processed meat containing 200 parts per million of sodium nitrite (a permissible residue) ingests enough of the compound to convert from 1.4 to 5.7 per cent of his hemoglobin to methemoglobin. Ordinarily, this percentage is insignificant. But if the same individual is a heavy smoker and lives in an air-polluted community, the cumulative loss in oxygen-carrying capacity produced by the nitrites in the food and by the carbon monoxide in tobacco smoke and motorcar exhaust can no longer be dismissed as trivial.

Sodium nitrite is highly toxic in relatively small amounts. About four grams of the compound constitutes a lethal dose for adults. Although Lehman regards 200 parts per million of sodium nitrite as a safe residue, he notes that "only a small margin of safety exists between the amount of nitrite that is safe and that which may be dangerous. The margin of safety is even more reduced when the smaller blood volume and the corresponding smaller quantity of hemoglobin in children is taken into account. This has been emphasized in the recent cases of nitrite poisoning in children who consumed weiners and bologna containing nitrite greatly in excess of the 200 ppm permitted by Federal regulations. The application of nitrite to other foods is not to be encouraged."

Emulsifiers have been used in bread not only as softening agents, which can give stale bread the texture of newly baked bread, but as substitutes for nourishing ingredients. "The record of the bread-standards hearings contain evidence of distribution among bakers of advertising material advocating the replacement of fats, oils, eggs and milk by emulsifiers," George L. Prichard, of the Department of Agriculture, told the Delaney Committee. "The use of such products as components of food may work to the disadvantage of our farm economy by displacing farm products normally used. The record indicates that natural food constituents, such as fats and oils, probably will be reduced in many commercial bakery products if bakers are allowed to employ these emulsifiers."

This substitution affects more than 117 the 117 farm 117 economy, however; it also works to the disadvantage of the consumer. To illustrate this, the Delaney Committee report compared the ingredients in two cake batters prepared by the same company eleven years apart; during the interval emulsifiers were added to the product. The first batter, made in 1939, did not contain synthetic emulsifiers; the second, prepared in 1949, did. "On a percentage basis, the cake batter in 1939 contained 13 per cent eggs and 8.6 per cent shortening.

In 1949, the cake batter contained 6.3 per cent eggs and 4.8 per cent shortening, with somewhat less than 0.3 per cent of synthetic emulsifier." The report noted that a "synthetic yellow dye could be added to provide the colour formerly obtainable through the use of eggs. The utilization of synthetic yellow dye in commercial cake was practiced before the war. There are indications that the use of artificial colouring matter is increased when quantities of whole eggs or egg yolks are reduced in commercial cake formulas." To heighten the insult, among the most commonly used emulsifiers in 1949 were the polyoxyethylene monostearates. The yellow dye referred to in the Delaney Committee report was probably FDC Yellow No. 3(Yellow AB), which, until fairly recently, was added to many yellow cake mixes. The dye often contained impurities of a potent carcinogen and its use in foods was forbidden in 1959. For a number of years, however, both the emulsifier and the dye undoubtedly appeared in many brands of cake and reached large numbers of unsuspecting consumers.

Problems of this kind are not likely to disappear unless there is a basic change in the viewpoint of the F.D.A. "Inherited from the Wiley era is a too common misconception that all 'chemicals' are harmful and the related idea that any amount of a 'poison' is harmful," the F.D.A. observes in a brochure on food additives. "The fact is, of course, that chemical additives, or food *additives* as they are now being called, have brought about great improvements in the American food supply. Additives like potassium iodide in salt and vitamins in enriched food products are making an important contribution to the health of our people and yet it is a fact that both iodine and some of the vitamins would be harmful if consumed in excessive amounts. Many similar examples could be given to refute these common misconceptions."

This argument is grossly misleading. Iodine and vitamins are indispensable to human life, but coal-tar dyes and benzoic acid, for example, are not. If we adhered to a well-balanced diet of properly prepared natural foods, iodine and vitamins would never enter the body in toxic amounts. Coal-tar dyes, on the other hand, are suspected of being harmful to man in nearly any amount if consumed repeatedly and the kindest statement that can be made for the presence of benzoic acid in food is that the compound is "safe under the conditions of its intended use."

The formula "safe under the conditions of its intended use" exposes the consumer to serious risks when it opens the door to the use of food additives whose biochemical activity is not understood. Many unexpected problems may arise when such additives appear in food. A particular additive may seem to be relatively harmless to the organs of the body, but it may be carcinogenic on the cellular level of life.

Another additive may produce insignificant or controllable effects when studied in isolation; brought into combination with various chemicals in food, however, it may give rise to toxic compounds. The body, in turn, may make a toxic additive more poisonous in the course of changing it during metabolism. "At one time it was generally believed that whenever a toxic substance was absorbed the body was capable of calling upon special mechanisms for detoxifying the toxicant," Lehman observes. "It was believed also that the metabolic pathway that the toxicant followed always proceeded in the direction of the formation of a less toxic compound.

Later work on the metabolism of drugs and toxic substances showed that special mechanisms did not exist, but that the toxicant was subject to the same metabolic influences as those which normally operate in the body. The assumption that the metabolic product was less poisonous than the parent substance from which it was derived is also unwarranted simply because in many instances the toxicity of either the original substance and its conversion product is unknown. In other instances the metabolic product is even more poisonous than its parent. The conversion product, heptachlorepoxide, which is two or three times more poisonous than the parent substance, heptachlor [a widely used insecticide], may be cited as an example of this."

Finally, food additives may cause allergic reactions that are likely to go unnoticed for many years. The reactions need not be severe to be harmful. Otto Saphir and his colleagues at the Michael Reese Hospital in Chicago have recently suggested that the development of arteriosclerosis may be promoted by the sensitivity of the body to allergenic compounds, notably certain antibiotics. By using sulfa drugs to produce allergies in forty-two rabbits, the researchers were able in eight months to cause degenerative changes in the arteries of thirty-one of the animals.

These changes closely resembled arteriosclerosis in man. According to a press account of Saphir's report, the rabbits' reactions "were not apparent on the surface." It would be very imprudent to assume that such effects are produced only by chemicals that cause severe or noticeable allergies. If the data of Saphir and his colleagues are applicable to man, additives with even minor allergenic properties cannot be dismissed as harmless.

Today more than 3,000 chemicals are used in the production and distribution of commercially prepared food. At least 1,288 are purposely added as preservatives, buffers, emulsifiers, neutralizing agents, sequestrants, stabilizers, anticaking agents, flavouring agents and colouring agents, while from 25 to 30 consist of nutritional supplements, such as potassium iodide and vitamins. The remaining chemicals are "indirect additives"—substances, such as detergents and germicides, that get into food by way of packaging materials and processing equipment. Many chemical additives are natural ingredients, but a large number are not. The artificial substances that appear in food range from simple inorganic chemicals to exotic compounds whose biochemical activity is still largely unknown.

MICROBIAL DETERMINATIONS

To assess the presence and extent of microbial contamination in food, it is standard practice to perform several types of microbial determinations. These determinations are important because microorganisms from foods can cause such diseases as staphylococcal food poisoning, salmonellosis, typhoid fever, cholera, and gastroenteritis.

THE STANDARD PLATE COUNT

One method for determining the number of bacteria in foods is the standard plate count. The procedure is performed by taking a gram of food sample and diluting it in 99-ml bottles of sterile water or buffer in the method described in the stage on aquatic microbiology. The number of bacterial colonies is multiplied by the dilution factor to determine the number of bacteria per gram of food.

COLIFORM AND FUNGAL DETERMINATIONS

In food, it is valuable to assess the number of coliform bacteria. These Gram-negative intestinal rods do not cause disease but are valuable indicators of fecal contamination. Presumably, when coliform bacteria are present, the food has been contaminated with fecal matter and is unfit to consume.

The coliform most commonly sought is Escherichia coli Various types of bacteriological media, such as violet red bile agar, are available for cultivating this bacteria, and the standard plate count technique can be performed to assess the number of coliform bacteria per gram of food. Fungi can be assessed in foods by using a medium such as Sabouraud dextrose agar.

This medium encourages fungal spores to germinate and form visible masses of filaments called mycelia. A count of the resulting mycelia gives an estimate of the fungal contamination of the food.

THE PHOSPHATASE TEST

To determine contamination in milk, the plate count technique can be used, but a more rapid test is the phosphatase test. Phosphatase is an enzyme destroyed by the pasteurization process. However, if the test for phosphatase shows that it is present after the milk has been treated, then the pasteurization has been unsuccessful. Testing for phosphatase is a more rapid and efficient method for determining contamination than the plate count technique.

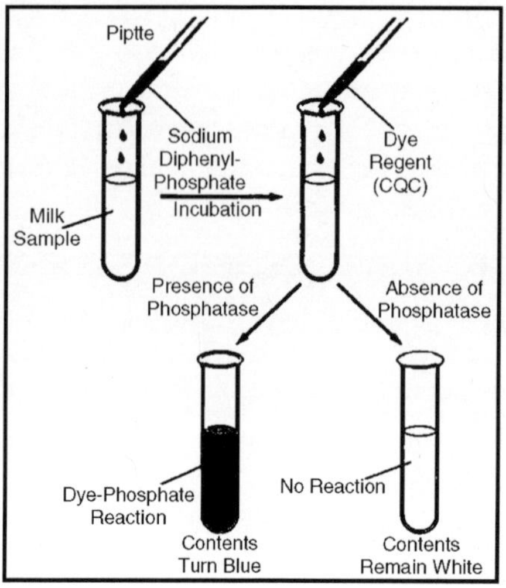

Fig. The Phosphatase Test

The substrate sodium diphenyl phosphate is added to a pasteurized milk sample, and the tube is incubated. Then the dye reagent is added. If phosphatase is present, the tube contents turn blue. However, they remain clear if phosphatase was destroyed during the pasteurization process.

FOODS FROM MICROORGANISMS

Microorganisms are widely used in the food industry to produce various types of foods that are both nutritious and preserved from spoilage because of their acid content.

DAIRY FOODS

In the dairy industry, many products result from fermentation by microorganisms in milk and the products of milk. For example, buttermilk results from the souring of low-fat milk by lactic acid. The flavour is due to substances such as diacetyl and acetaldehyde, which are produced by species of Streptococcus, Leuconostoc, and Lactobacillus as they grow.

A fermented milk product with a puddinglike consistency is yogurt. Two bacteria, Streptococcus thermophilus and Lactobacillus bulgaricus, are essential to its production. After the milk has been heated to achieve evaporation, the bacteria are added, and the condensed milk is set aside at a warm temperature to produce the yogurt.

Sour cream is produced in a similar way, using cream as a starter material. The protein portion of the milk, the casein, is used to produce cheese and cheese products. Precipitated from the milk, the protein curd is an unripened cheese such as cottage cheese. The leftover liquid, the whey, can be used to make cheese foods. When the cheese is allowed to ripen through the activity of various microorganisms, various cheeses are produced.

Soft cheeses, such as Camembert, do not spoil rapidly. Camembert cheese is a product of the growth of the fungus Penicillium camemberti. Hard cheeses have less water and are ripened with bacteria or fungi. Swiss cheese is ripened by various bacteria, including species of Propionibacterium, which produces gas holes in the cheese. Bleu cheese is produced by Penicillium roqueforti, which produces veins within the cheese as it grows.

OTHER FERMENTED FOODS

Other fermented foods are also the product of microbial action. Sauerkraut, for example is produced by *Leuconostoc* and *Lactobacillus* species growing within shredded cabbage. Cucumbers are fermented by these same microorganisms to produce pickles.

BREAD

Bread is still another product of microbial action. Flour, water, salt, and yeast are used to make the dough. The yeast most often used is *Saccharomyces cerevisiae*. This organism ferments the carbohydrates in the dough and produces carbon dioxide, which causes the dough to rise and creates the soft texture of bread. Unleavened bread is bread that contains no yeast. Sourdough bread can be made by using lactic acid bacteria to contribute a sour flavour to the dough.

GM FOOD AND FOOD SECURITY

WHAT IS FOOD SECURITY

The official definition of food security, adopted at the World Food Summit of 1996, states:

- "Food security exists when all people, at all times, have physical and economic access to sufficient, safe and nutritious food to meet their dietary needs and food preferences for an active and healthy life."

This definition is understood within the framework of sustainability and was drawn from chapter 14.6 of Agenda 21, adopted at the 1992 United Nations Conference on Environment and Development, which states: "The major thrust of food security...is to bring about a significant increase in agricultural production in a sustainable way and to achieve a substantial improvement in people's entitlement to adequate food and culturally appropriate food supplies." The underlying assumption is that the means of increasing food availability in many countries exist, but are not being realised because of a range of constraints. In identifying and resolving these constraints, it is necessary to find sustainable ways of improving and reducing year-to-year variability in food production and open the way for broader food access.

The causes of food insecurity involve a complex interplay of economic, social, political and technical issues. An analysis of this interplay should determine the potential solution and best approach for a given population group. The issue for some communities is being able to produce sufficient food. For others, lack of money to purchase a wider selection of foods is the problem. Food insecurity and poverty are strongly correlated. The Swedish International Cooperation Agency defines poverty as a three-fold deficiency: a lack of security, ability and opportunity. Poverty is the main cause of food insecurity, and hunger is also a significant cause of poverty. Hunger is not only about quantity—it goes hand-in-hand with malnutrition. Food insecurity and malnutrition impair people's ability to develop skills and reduce their productivity.

A lag in farm productivity is closely associated with rural poverty and hunger. Food insecurity nevertheless is a reality experienced by the vulnerable in all societies and in all countries, developed and developing. In developed countries, the problem of food security is often a reflection of affordability and accessibility through conventional channels. Food security for the rural poor in developing countries is about producing or securing enough to feed one's household and being able to maintain that level of production year after year.

Hunger and malnutrition increase susceptibility to disease and reduce people's ability to earn a livelihood. In instances where hunger is related to household income, improving food security by ensuring access to food or increasing the purchasing power of a family is essential. Providing poor communities with the skills to improve conditions in an economically and ecologically sustainable manner creates a window of opportunity to alleviate poverty at the subsistence-farming level and, on a larger scale, by having an impact on the economic development of the country.

THE CHALLENGES TO FOOD SECURITY

In developing countries, 800 million people are undernourished, of which a significant proportion live on less than US$1 per day, despite a more than 50% decline in world food prices over the past 20 years. Global food production has soared, making a variety of foods available to all consumers. Although the decline in food prices in developed countries has benefited the poor who spend a considerable share of their income on food, this trend has not had much impact on the majority in the developing world, with sub-Saharan Africa painting the gloomiest picture.

Due to the substantial price reduction in this commodity sector, cereals have become the staple foods in the diet of poor people. While yield increases in the major cereal crops has meant more calorific intake of food, micronutrient malnutrition remains a serious problem. Regional analyses depict sub-Saharan Africa as the only region where both the number and proportion of malnourished children have consistently risen in the past three decades. However, malnutrition in South Asia is also very high.

The world population is projected to reach 8 billion by 2025, and it is estimated that most of this growth will occur in developing countries. Feeding and housing an additional 2 billion people will cause considerable pressure on land, water, energy and other natural resources. Looking at projections to 2020, the worldwide per capita availability of food is expected to increase by approximately 7%, *i.e.* 2900 calories per person per day.

Nonetheless, an average availability of 2300 calories is projected for individuals in sub-Saharan Africa, a figure that is just above the recommended minimum calorie intake for an active and productive life. In terms of agricultural output, preliminary world estimates for 2001 suggested that growth was as low as 0.6%. Annual rates also demonstrate a trend of decreased productivity, particularly in developing-country regions. Output growth in Asia has systematically declined over the past five years the rates experienced in sub-Saharan Africa are lower than average. Agricultural productivity is important for food security in that it has an impact on food supplies, prices, and the incomes and purchasing power of farmers. Improving food security at the national level requires an increase in the availability of food through increased agricultural production, or by increasing imports.

To augment domestic production and maintain an adequate supply of food, food-insecure countries often rely on imports and food aid. Export earnings are frequently low and do not suffice to provide foreign exchange to finance imports. Thus, in the long term, importing food is unsustainable. Historically, increased food production in the developing countries can be attributed to the cultivation of more land rather than to the deployment of improved farming practices or to the application of new technologies. By its very nature, agriculture threatens other ecosystems, a situation that can be exacerbated by over-cultivation, overgrazing, deforestation and bad irrigation practices. However, increased demands for food in Asia, Europe and North Africa have to be met by increasing yields because most land in these areas is already used for agriculture.

The potential to expand agricultural land exists in Latin America and sub-Saharan Africa only, where much of the remaining land is marginal for agricultural expansion. The implication, therefore, is that the increase in food production needed to feed the world's growing population can only be met by increasing the amount of food produced per hectare. Recognizing the extent of environmental degradation caused mainly by human activities, the multilateral agreements that arose from the UNCED meeting of 1992 were intended to address the compromised food-security situation on a global scale. One such agreement is the United Nations Convention to Combat Desertification.

This agreement promotes the implementation of practices intended to reverse desertification for sustainable land use and food security. As the more affluent developed countries tend to produce more food, some argue that redistribution of these surpluses could feed the escalating populations of developing countries.

Redistribution, however, requires policy changes that may be impossible to implement on a global scale. Therefore, a substantial proportion of the food demands of developing countries will have to be met by the agricultural systems in these countries. Enabling a consistent and sustainable supply of food will require an overhaul in the production processes and the supporting infrastructure. Finding solutions to declining crop yields requires an effort that will improve the assets on which agriculture relies; namely, soils, water and biodiversity.

Transforming the agricultural systems of rural farmers by introducing technologies that integrate agro-ecological processes in food production, while minimizing adverse effects to the environment, is key to sustainable agriculture. In addition, increases in crop yield must be met with the use of locally available low-cost technologies and minimum inputs without causing damage to the environment.

ATTAINING FOOD SECURITY

Within the context of the definition, three distinct components appear central to attaining food security: availability, accessibility and adequacy. Within each component, questions are raised which may need to be addressed to improve the food security situation at a national, regional or international level.

The questions raised here are intended to demonstrate the complexity of the issue and are by no means exhaustive:

- *Availability*: is there enough food available through domestic production or imports to meet the immediate needs? Is production environmentally sustainable to meet long-term demands? Are the distribution systems effective in reaching low-income and rural communities?
- *Accessibility*: do the vulnerable in society have the purchasing power to attain food security? Can they afford the minimum basic diet of 2100 calories per day required for an active and productive life?
- Adequacy: does the food supply provide for the differing nutritional needs, *i.e.* a balanced diet, offering the necessary variety of foods at all times? Is the food properly processed, stored and prepared?

Global food productivity is undergoing a process of rapid transformation as a result of technological progress in the fields of communication, information, transport and modern biotechnology. A general observation is that technologies tend to be developed in response to market pressures, and not to the needs of the poor who have no purchasing power. As agriculture is the main economic activity of rural communities, optimizing the levels of production will generate employment and income, and thus uplift the wealth and well-being of the community.

Improving agricultural production in developing countries is fundamental to reducing poverty and increasing food security. Investment to raise agricultural productivity can be achieved through the introduction of superior technologies such as better-quality seeds, crop rotation systems etc. It is argued, however, that the adoption of earlier agricultural technologies has led to the emergence of more virulent strains of pests, pathogens and weeds, soil deterioration and a loss of biodiversity.

The Green Revolution, in particular, focused on wheat and rice—not much attention was paid to staple crops such as sorghum, cassava or millet. Also, the seeds and fertilizers required to grow the higher-yielding varieties were expensive and therefore not accessible to all. Reaffirming support for the principles agreed upon at the UNCED, the United Nations Millennium Development Goals have set a road map for protecting the environment. Embraced in these time-bound goals is a new development ethic that demands sustainability in a framework where progress is measured in terms of actions reconciling the economic and ecological factors of food production for the benefit of present and future generations?

Extending this understanding to agriculture, sustainable agriculture is defined as:

- Environmentally sound, preserving resources and maintaining production potential;
- Profitable for farmers and workable on a long-term basis;
- Providing food quality and sufficiency for all people;
- Socially acceptable; and
- Socially equitable, between different countries and within each country.

A secure food system is one in which the ecological resources on which food production depends allow for their continued use, with minimum damage for present and future generations.

In other words, food security and sustainable agriculture are interlinked and both are central to the concept of sustainable development. The FAO Anti-hunger Programme reported that increasing investment in agriculture and rural development can reduce hunger. To reduce the number of hungry people by half by 2015, it estimated that funding of US$24 billion would be required for agricultural research, emergency food assistance, and improving rural infrastructure. By contrast, at the current rate of progress, the number of food-insecure would fall by only 24%. For people to be food-secure, they must have access to the resources needed to buy or produce their own food.

Breaking the poverty cycle of rural communities, whose livelihoods depend on agriculture, will require investment in different technologies to address the different constraints experienced in different world regions. The production problems experienced by farmers vary between countries and communities, and technological solutions need to be relevant to those circumstances, *i.e.* one solution will not be suitable everywhere.

The potential of some of these technologies has been demonstrated in various world regions— for instance, agro-ecological improvement programmes involving:

- Better harvesting and conservation of water, even in rain-fed environments;
- A reduction of soil erosion by adopting zero tillage combined with the use of green manure and herbicides as in Argentina and Brazil; and
- Pest and weed control without pesticide or herbicide use, *e.g.* Bangladesh and Kenya, has been well tested and established.

Indeed, such programmes are now widely accepted as being at the core of sustainable agriculture. The communities that participated in these projects were able to transform food production through the use of resource-management strategies that focused on improving the soil by growing leguminous crops and applying agro-forestry, zero tillage and green manure. These and other projects have proved that the sustainability of any farming practice and the conditions under which production can be maintained at reasonable levels cannot be predicted with absolute certainty. Some regions may be better able to transfer high-yielding technologies with varying degrees of success.

The uptake of new production systems has proved successful where the programmes have included the participation of entire communities and have not been introduced to isolated groups of farmers. Producing nutritionally enhanced properties in staple crops eaten by the poor could reduce the burden of disease in many developing countries.

Scientists at the International Crops Research Institute for the Semi-Arid Tropics have developed a pearl millet variety enhanced with beta-carotene. The trait naturally occurs in two Burkina Faso millet lines from which it was transferred by conventional breeding methods. Genetic modification of japonica rice with a ferritin gene has not given superior results compared to rice with an 80% increase in iron density produced by conventional plant breeding at the International Rice Research Institute.

Research and technology alone will not drive agricultural growth. Inadequate infrastructure and poorly functioning markets tend to exacerbate the problem of food insecurity. The cost of marketing farm produce can be prohibitive for small-scale farmers, as their isolation prevents the link between agricultural and non-agricultural activities among adjacent villages and between rural and urban areas.

Building roads in rural areas is vital to facilitating growth, trade and exchange of farm and non-farm products in rural communities, even those that can adequately feed themselves. For instance, government investment in irrigation projects, storage and transport facilities, roads connecting villages to larger markets in the rural areas of China and India, has made an impressive impact on employment and productivity and ultimately provided opportunities for poverty alleviation in the affected areas. UNDP, basic thresholds in roads, power, ports and communications must be reached in order to sustain growth.

A POTENTIAL ROLE FOR MODERN BIOTECHNOLOGY

The Convention on Biological Diversity dictates the use and application of relevant technologies as a means of achieving the objectives of conservation and sustainable use with specific reference to biotechnology. Modern biotechnology is purported, from a technical perspective, to have a number of products for addressing certain food-security problems of developing countries. It offers the possibility of an agricultural system that is more reliant on biological processes rather than chemical applications.

The potential uses of modern biotechnology in agriculture include: increasing yields while reducing inputs of fertilizers, herbicides and insecticides; conferring drought or salt tolerance on crop plants; increasing shelf-life; reducing postharvest losses; increasing the nutrient content of produce; and delivering vaccines. The availability of such products could not only have an important role in reducing hunger and increasing food security, but also have the potential to address some of the health problems of the developing world.

Achieving the improvements in crop yields expected in developing countries can help to alleviate poverty: directly by increasing the household incomes of small farmers who adopt these technologies; and indirectly, through spill-overs, as evidenced in the price slumps of herbicides and insecticides. Indirect benefits as a whole tend to have an impact on both technology adopters and non-adopters, the rural and urban poor. Indeed, some developing countries have identified priority areas such as tolerances to alkaline earth metals, drought and soil salinity, disease resistance, crop yields and nutritionally enhanced crops.

The adoption of technologies designed to prolong shelf-life could be valuable in helping to reduce postharvest losses in regionally important crops. Prime candidates in terms of crops of choice for development are the so-called 'orphan crops', such as cassava, sweet potato, millet, sorghum and yam. Multinationals have found no incentive to develop these crops and have instead invested in marketable crops with high profit returns. This strategy is intended to target wealthier farmers in temperate-zone countries with the financial capacity and tradition of supporting new seed products. However, here is a potential for multinational companies to develop crops grown largely in developing countries. The investment costs are low and the potential markets considerably large. While some public-sector research institutes in developing countries are forging ahead with the application of modern biotechnology, a small number are supported by government policy and therefore follow a defined agenda.

Still other governments believe that the risks associated with modern biotechnology outweigh the benefits. Currently, the many promises of modern biotechnology that could have an impact on food security have not been realised in most developing countries. In fact, the uptake of modern biotechnology has been remarkably low owing to the number of factors that underpin food security issues.

In part, this could be because the first generation of commercially available crops using modern biotechnology were modified with single genes to impart agronomic properties with traits for pest and weed control, and not complex characteristics that would modify the growth of crops in harsh conditions. Secondly, the technologies are developed by companies in industrialized countries with little or no direct investment in, and which derive little economic benefit from, developing countries. Thirdly, many developing countries do not have the necessary biosafety frameworks to regulate the products of modern biotechnology.

For example, it took over two years for the Kenyan authorities to approve the field-testing of a virus-resistant sweet-potato variety because the scientific capacity for evaluating the product was unavailable. It should be noted, however, that such delays in the approval process have also been seen in developed countries, especially during the initiation of national regulatory evaluation. However, this trend is quickly changing as a number of developing countries either adopt or develop appropriate biotechnologies or regulatory infrastructures.

A report by the International Service for National Agricultural Research states that more than 40 crops are the focus of public-sector research programmes of 15 developing countries involving disease-resistant traits in rice, potato, maize, soybean, tomato, banana, papaya, sugarcane, alfalfa and plantain. For example, the Brazilian Agricultural Research Corporation has concentrated research into genetically modified crops on disease resistance in beans, papaya and potatoes.

The research programme at the University of Cape Town focuses on the development of crops resistant to viruses and to desiccation. The university has recently had a breakthrough with maize streak-virus resistance. In Thailand, the National Centre for Genetic Engineering and Biotechnology has supported research into disease resistance of rice, pepper and yard-long beans. therefore follow a defined agenda. Still other governments believe that the risks associated with modern biotechnology outweigh the benefits. Currently, the many promises of modern biotechnology that could have an impact on food security have not been realised in most developing countries. In fact, the uptake of modern biotechnology has been remarkably low owing to the number of factors that underpin food security issues.

In part, this could be because the first generation of commercially available crops using modern biotechnology were modified with single genes to impart agronomic properties with traits for pest and weed control, and not complex characteristics that would modify the growth of crops in harsh conditions. Secondly, the technologies are developed by companies in industrialized countries with little or no direct investment in, and which derive little economic benefit from, developing countries. Thirdly, many developing countries do not have the necessary biosafety frameworks to regulate the products of modern biotechnology.

For example, it took over two years for the Kenyan authorities to approve the field-testing of a virus-resistant sweet-potato variety because the scientific capacity for evaluating the product was unavailable. It should be noted, however, that such delays in the approval process have also been seen in developed countries, especially during the initiation of national regulatory evaluation. However, this trend is quickly changing as a number of developing countries either adopt or develop appropriate biotechnologies or regulatory infrastructures.

A report by the International Service for National Agricultural Research states that more than 40 crops are the focus of public-sector research programmes of 15 developing countries involving disease-resistant traits in rice, potato, maize, soybean, tomato, banana, papaya, sugarcane, alfalfa and plantain. For example, the Brazilian Agricultural Research Corporation has concentrated research into genetically modified crops on disease resistance in beans, papaya and potatoes. The research programme at the University of Cape Town focuses on the development of crops resistant to viruses and to desiccation. The university has recently had a breakthrough with maize streak-virus resistance. In Thailand, the National Centre for Genetic Engineering and Biotechnology has supported research into disease resistance of rice, pepper and yard-long beans.

Developing countries with limited financial and human resources need to find the right balance for investing in conventional and modern biotechnology research programmes. While alliances with the private sector may contribute to the search for new technologies, the public sector needs to focus on crops and traits in which the former may be unwilling or unable to invest. The extent to which priority is given to modern biotechnology over other research methods should be linked to a country's agricultural priorities and objectives as well as to its environmental concerns. Ultimately, investment in interventions that support good governance, the development of rural infrastructure and market access is required before any of the promises of modern biotechnology can be realised. In general, policies that stimulate economic growth and target poverty reduction may have significant bearing on the health and well-being of the population.

RESEARCH OWNERSHIP

Research is a critical part of any effort aimed at improving food production and reducing poverty. Globally, much of agricultural R&D is carried out in the public sector, thereby serving the interests of developing countries. Public research in developed countries and Latin America is mostly conducted by government institutions and universities, whereas almost all agricultural research in Africa is carried out in public institutions, including R&D, technology transfer and dissemination of improved plant varieties. In general, international agricultural research institutions form a second level of research development and technology providers in developing countries.

Public research institutes have, in the past, researched and improved orphan crops, mainly for donation to poor farmers or at cost. Generally, academic institutions are perceived as producers of knowledge that benefit and protect the public. Also, national and international research institutes aim to address the agricultural problems of resource-poor farmers in developing countries, *e.g.* increasing productivity through the use of a variety of techniques, including modern biotechnology.

In reality, public institutions are now exposed to the forces of globalization and compelled to compete for their survival. In the current climate, government intervention in R&D worldwide has dwindled; hampering the level of innovation generated for public good. In fact, research facilities in many developing countries are poorly equipped, often limiting experiments to traditional and outdated research.

The diminishing role of public research institutes is perceived to have a major impact on the adoption of modern biotechnology in terms of introducing relevant products to those that need them most. Most of the field trials in the EU and the USA are conducted by private companies. An analysis of field-trial data from the USA shows that three crops account for 64% of all trials, of which 69% express herbicide- and pest-resistance traits. Of the trials conducted in the EU, 67% involve maize, sugar beet and rapeseed, and 71% of the novel gene categories presented herbicide- or pest-resistance traits. Less than 1% of all the trials in the EU and the USA are of plant varieties grown in tropical and subtropical climates, half of which have been conducted by the public sector.

Most of the public-sector research involving modern biotechnology in developing countries is still in the laboratory phase - none of the crops have progressed to marketable products. China, on the other hand, has approved the field-testing of over 500 GMOs to date and the commercial release of 50, including a prolonged shelf-life tomato, virus-resistant sweet pepper and vaccines for animal use.

The experiences limiting the progression to commercializing research efforts range from: a lack of resources for meeting the high costs of regulatory requirements; lack of foresight, planning and business acumen for enabling the transition from research to a commercial product; and lack of capacity to negotiate patent licenses. Also, developments in modern biotechnology have occurred independently of the sustainable agricultural goals and priorities of the developing countries concerned. Furthermore, a needs assessment for a particular technology has often not been carried out before a research project is begun.

Nevertheless, it is often argued that the commercialization of some products would encourage monocropping as national agricultural research has focused on a few crops, whereas rural communities tend to grow a wide range of crop species and plant varieties. The focus of international agricultural research centres is on plant production and protection, livestock production and health and food processing. With respect to food crops, research emphasis appears to be spread equally between cereals, root crops and legumes. However, within the cereals, research devoted to rice far outweighs research on maize and sorghum.

A large proportion of the total agricultural research activities of many developing countries are donor-funded. International research institutes, such as the Consultative Group on International Agricultural Research, depend on government grants and donations from philanthropic organizations for their survival, and yet investment in this sector has fallen in real terms. A significant fraction of the funding spent on international agricultural research institutes is used on activities covering a relatively large number of crops.

The beneficiaries of such initiatives are a small number of countries with relatively advanced scientific capabilities. During the 1990s, developing countries as a group invested more in agricultural research than developed countries, even though the spending was unevenly distributed. In industrialized countries, private-sector investment in R&D far exceeds government spending on technology development, so that much of the public good previously entrusted to public research institutes is now privately owned. In comparison, private-sector investment in developing countries is around 1% of total global spending in this sector and developing countries invest less than 5% of the total private sector spending in biotechnology. Although the agricultural sector in developing countries is large and of significant importance to the domestic economy, spending in agricultural research does not match this level of activity. For instance, 80% of the food consumed in sub-Saharan Africa is obtained from domestic production.

Impact of Intellectual Property Rights on Research

Intellectual property rights have been relevant to agriculture since their inception but have gained importance with regard to research in developed countries in the past 20-30 years. In particular, IPRs have been used to protect and preserve the value of products produced by conventional methods, *e.g.* the trademark registration of food products.

The rationale for IPRs is that they encourage the inventor to advertise the invention and disclose the new knowledge, while simultaneously holding the rights to protect the invention from competitors. Disseminating this information is thought to stimulate new ideas and further rounds of innovation and technological advancement.

IPRs afford time-limited protection to artistic, scientific, technological or economic products, and can be protected by way of copyrights, trademarks, design patents, utility patents, plant patents, plant breeders' rights and trade-secret law. Of these mechanisms, patents are considered the most powerful tool of the IPR system. Patents play different roles in different technologies and sectors. Patent protection of biotechnology makes it a tool for technology transfer and securing new markets in a global economy. Without protection, new ideas and information are entirely in the public domain.

This can, in certain systems, lead to underinvestment in R&D or the withholding of knowledge. Plant variety protection provides less protection than patents in that generally it makes provision for farmers' rights, allowing them to use harvested seed, and includes an exemption for research use. Despite the increase in availability, new plant varieties continue to be inaccessible or inappropriate for poor farmers, and the rate of innovation remains largely unchanged in countries with a PVP system.

Studies have indeed shown that in middle-income countries, the principal beneficiaries of PVPs are commercial farmers and the seed industry. PVP is seen as a system that protects small advances in plant breeding, while a patent regime is thought to lead to the protection of big leaps in technological achievements. Patent protection for products of modern biotechnology is important because they are expensive to develop and easy to copy. Even so, developing countries have limited capabilities to innovate in industrial fields such as modern biotechnology, and to effectively enforce IPRs. A significant number of developing countries have not established intellectual property regimes that cover plants.

This situation may thus discourage private-sector investment. With no assurance that they can recoup some profit on GM products, multinationals are unlikely to devote much attention to the challenges of developing countries unless seen in a development aid context or through public-private partnership. Although this situation impedes private-sector investment in developing countries, it also implies that the freedom to operate on products designated for local markets is not hindered. Exercising this freedom to operate is not a well-understood concept. For example, in the case of 'golden rice', where permission was required for the use of about 70 patents, the impression was that the patents were being relinquished in favour of the poor. In fact, most of the patents involved are not valid in the major rice-consuming countries.

The technology donated for the development of virus-resistance in non-commercial potato varieties is free of patents relevant to Mexico, and the same holds true in the case of virus-resistant sweet potatoes in Kenya. Researchers are usually unaware of the proprietary status of the technologies they are using in their work. Nevertheless, the proliferation of broad patents is thought to impede the research capabilities of other interested parties. Some countries grant very broad patents conferring monopoly rights over large areas of research, thereby potentially threatening the other goal of intellectual property, namely the right to build upon the original invention. The prevailing patent rules have the potential to limit the accessibility of these technologies to public institutions and ultimately poor farmers.

Furthermore, the strengthening of IPRs is thought to restrict the flow of germplasm and inhibit the development of new plant varieties. This is because if and when researchers in public institutions do get permission to develop the technologies further, access is granted under licence agreements with restrictions on commercializing innovations. It is also argued that a stringent, multilateral IPR system will not benefit all countries equally. Indeed, the benefits will largely be influenced by the economic and technological levels of development in each country.

The United Kingdom Commission on Intellectual Property Rights, "the critical issue in respect of IPRs is perhaps not whether it promotes trade or foreign investment, but how it helps or hinders developing countries to gain access to technologies that are required for their development." During the 1990s, obtaining a patent application in the USA cost US$20,000 and twice as much in the EU. In general, PVP is cheaper; valued at one-tenth of the patent price. Moreover, the preparation of a food-safety dossier for a product derived from modern biotechnology, for example, is estimated to be around US$1 million.

These estimates cannot be compared with the regulatory costs in developing countries. The cost of regulation in developing countries does not encourage the commercialization of products of modern biotechnology developed by public-sector research institutes. In most cases, the regulatory costs far exceed the research costs.

Many developing countries do not have the resources to match private-sector investment in modern biotechnology. In this new playing field, public institutions also need resources to deal with intellectual property rights to help compensate and increase the public benefit. Otherwise their involvement in R&D could be deterred by lack of funding. If public institutions are to use the techniques of modern biotechnology, then the use of IPRs as a framework for facilitating technology transfer must be emphasized more than its handling as a revenue-generating system. IPRs can, however, play a major role in clarifying the mechanisms for access to technology, and determine the downstream aspects of use and exploitation of genetic resources.

There are several ways in which public institutions and small companies in developing countries can gain access to patented genes and enabling technologies to overcome the current barriers to research. The first of these includes a measure of goodwill by multinationals to relinquish their rights to technologies for use by researchers in developing countries by adopting programmes of social responsibility as in the case of 'golden rice', a variety containing beta-carotene, virus-resistant sweet potatoes and a virus-resistant non-commercial potato variety in Mexico.

A different type of programme initiated in the USA has established an intellectual property clearing house to make information on the intellectual property owned by public research institutes, including universities, available to researchers around the world.

There are also suggestions that redesigning patent laws to narrow the type and scope of patent coverage ought to make more technologies accessible to public institutions. The thinking behind these suggestions is that applying a stronger standard for rejecting patent applications for inventions that are 'obvious' should deter the patenting of minor inventions. In addition, a law that requires an invention to be genuinely useful in theory should reduce the number of patent applications being submitted. At present, it is possible in some countries to submit patent applications for abstract concepts that potentially protect large areas of research and thereby exclude innovation by others. Another option that may be attractive to developing countries is the creation of collaborations that involve research institutes, universities and the private sector.

The nature of these collaborations is likely to be influenced by the level of expertise and resources within the national public research institutes. Where a solid knowledge base exists, the public partners may be in a position to develop or acquire a technology that could be transferred into locally adapted varieties. Smaller institutes are more likely to provide the genetic resources and a positive public image. It is believed that such alliances would benefit public institutions and private companies; offering them an opportunity to license and distribute the technology.

The most well-known public-private sector partnerships include organizations such as the International Service for the Acquisition of Agri-biotech Applications that negotiate access to private-sector technologies for the improvement of subsistence crops and/or the transfer of technology and know-how. Although several types of public-private sector alliances already exist, the two newly established initiatives worth mentioning are the Global Cassava Partnership and the African Agricultural Technology Foundation, launched in November 2002 by FAO and in March 2003 by the Rockefeller Foundation, respectively. The former is a partnership involving some of the world's leading experts in cassava research, working mainly in public institutions. The AATF intends to function as a clearing house of available technologies with the primary aim of improving food security and reducing the poverty situation of small farmers, by facilitating transfer and use of appropriate technologies.

Such licensing arrangements have been put to the test in other fields. However, as in the case of the AATF, a clearing house is required to acquire the necessary technologies and permit their further use for developing-country needs. The drawback may be a requirement to divide the commercial sector into subsistence, middle-income and commercial markets. This market division may not be easy to achieve as some large developing countries have both commercially important markets and subsistence farmers.

A briefing document commissioned by the United Nations Industrial Development Organization proposes six activities that build upon private-sector investment and enable biotechnology transfer:

- Enabling government policies;
- Access to up-to-date authoritative information;
- Regional brokering service to strengthen public-private partnerships;
- A regional biotechnology investment service;
- An international intellectual property escrow service; and
- Initiatives for risk-shifting.

Each of the proposals can be implemented as a stand-alone project or in combination, as best suits the national and/or regional situation.

CLIMATE CHANGE AND FOOD SECURITY

IMPACTS ON FOOD PRODUCTION AND AVAILABILITY

Climate change affects agriculture and food production in complex ways. It affects food production directly through changes in agro-ecological conditions and indirectly by affecting growth and distribution of incomes, and thus demand for agricultural produce. Impacts have been quantified in numerous studies and under various sets of assumptions. A selection of these results is presented in *Quantifying the Impacts on Food Security*. Here it is useful to summarize the main alterations in the agro-ecological environment that are associated with climate change.

Changes in temperature and precipitation associated with continued emissions of greenhouse gases will bring changes in land suitability and crop yields. In particular, the Intergovernmental Panel on Climate Change (IPCC) considers four families of socio-economic development and associated emission scenarios, known as Special Report on Emissions Scenarios (SRES) A2, B2, A1, and B1. Of relevance to this review, of the SRES scenarios, A1, the "business-as-usual scenario," corresponds to the highest emissions, and B1 corresponds to the lowest. The other scenarios are intermediate between these two. Importantly for agriculture and world food supply, SRES A2 assumes the highest projected population growth of the four (United Nations high projection) and is thus associated to the highest food demand.

Depending on the SRES emission scenario and climate models considered, global mean surface temperature is projected to rise in a range from 1.8°C (with a range from 1.1°C to 2.9°C for SRES B1) to 4.0°C (with a range from 2.4°C to 6.4°C for A1) by 2100. In temperate latitudes, higher temperatures are expected to bring predominantly benefits to agriculture: the areas potentially suitable for cropping will expand, the length of the growing period will increase, and crop yields may rise.

A moderate incremental warming in some humid and temperate grasslands may increase pasture productivity and reduce the need for housing and for compound feed. These gains have to be set against an increased frequency of extreme events, for instance, heat waves and droughts in the Mediterranean region or increased heavy precipitation events and flooding in temperate regions, including the possibility of increased coastal storms; they also have to be set against the fact that semiarid and arid pastures are likely to see reduced livestock productivity and increased livestock mortality.

In drier areas, climate models predict increased evapotranspiration and lower soil moisture levels. As a result, some cultivated areas may become unsuitable for cropping and some tropical grassland may become increasingly arid. Temperature rise will also expand the range of many agricultural pests and increase the ability of pest populations to survive the winter and attack spring crops.

Another important change for agriculture is the increase in atmospheric carbon dioxide (CO_2) concentrations. Depending on the SRES emission scenario, the atmospheric CO_2 concentration is projected to increase from H"379 ppm today to >550 ppm by 2100 in SRES B1 to >800 ppm in SRES A1FI. Higher CO_2 concentrations will have a positive effect on many crops, enhancing biomass accumulation and final yield. However, the magnitude of this effect is less clear, with important differences depending on management type (*e.g.*, irrigation and fertilization regimes) and crop type. Experimental yield response to elevated CO_2 show that under optimal growth conditions, crop yields increase at 550 ppm CO_2 in the range of 10 per cent to 20 per cent for C_3 crops (such as wheat, rice, and soybean), and only 0–10 per cent for C_4 crops such as maize and sorghum. Yet the nutritional quality of agricultural produce may not increase in line with higher yields. Some cereal and forage crops, for example, show lower protein concentrations under elevated CO_2 conditions.

Finally, a number of recent studies have estimated the likely changes in land suitability, potential yields, and agricultural production on the current suite of crops and cultivars available today. Therefore, these estimates implicitly include adaptation using available management techniques and crops, but excluding new cultivars from breeding or biotechnology. These studies are in essence based on the FAO/International Institute for Applied Systems Analysis (IIASA) agro-ecological zone (AEZ) methodology.

For instance, pioneering work in ref. 9 suggested that total land and total prime land would remain virtually unchanged at the current levels of 2,600 million and 2,000 million hectares (ha), respectively. The same study also showed, however, more pronounced regional shifts, with a considerable increase in suitable cropland at higher latitudes (developed countries +160 million ha) and a corresponding decline of potential cropland at lower latitudes (developing countries "110 million ha).

An even more pronounced shift within the quality of cropland is predicted in developing countries. The net decline of 110 million ha is the result of a massive decline in agricultural prime land of H"135 million ha, which is offset by an increase in moderately suitable land of >20 million ha. This quality shift is also reflected in the shift in land suitable for multiple cropping. In sub-Saharan Africa alone, land for double cropping would decline by between 10 million and 20 million ha, and land suitable for triple copping would decline by 5 million to 10 million ha.

At a regional level, similar approaches indicate that under climate change, the biggest losses in suitable cropland are likely to be in Africa, whereas the largest expansion of suitable cropland is in the Russian Federation and Central Asia.

Impacts on the Stability of Food Supplies

Global and regional weather conditions are also expected to become more variable than at present, with increases in the frequency and severity of extreme events such as cyclones, floods, hailstorms, and droughts. By bringing greater fluctuations in crop yields and local food supplies and higher risks of landslides and erosion damage, they can adversely affect the stability of food supplies and thus food security.

Neither climate change nor short-term climate variability and associated adaptation are new phenomena in agriculture, of course. As shown, for instance, in ref. 9, some important agricultural areas of the world like the Midwest of the United States, the northeast of Argentina, southern Africa, or southeast Australia have traditionally experienced higher climate variability than other regions such as central Africa or Europe.

They also show that the extent of short-term fluctuations has changed over longer periods of time. In developed countries, for instance, short-term climate variability increased from 1931 to 1960 as compared with 1901 to 1930, but decreased strongly in the period from 1961 to 1990. What is new, however, is the fact that the areas subject to high climate variability are likely to expand, whereas the extent of short-term climate variability is likely to increase across all regions. Furthermore, the rates and levels of projected warming may exceed in some regions the historical experience.

If climate fluctuations become more pronounced and more widespread, droughts and floods, the dominant causes of short-term fluctuations in food production in semiarid and subhumid areas, will become more severe and more frequent. In semiarid areas, droughts can dramatically reduce crop yields and livestock numbers and productivity.

Again, most of this land is in sub-Saharan Africa and parts of South Asia, meaning that the poorest regions with the highest level of chronic undernourishment will also be exposed to the highest degree of instability in food production. How strongly these impacts will be felt will crucially depend on whether such fluctuations can be countered by investments in irrigation, better storage facilities, or higher food imports. In addition, a policy environment that fosters freer trade and promotes investments in transportation, communications, and irrigation infrastructure can help address these challenges early on.

Impacts of Climate Change on Food Utilization

Climate change will also affect the ability of individuals to use food effectively by altering the conditions for food safety and changing the disease pressure from vector, water, and food-borne diseases. The IPPC Working Group II provides a detailed account of the health impacts of climate change of its fourth assessment report. It examines how the various forms of diseases, including vector-borne diseases such as malaria, are likely to spread or recede with climate change. This chapter focuses on a narrow selection of diseases that affect food safety directly, *i.e.*, food and water-borne diseases.

The main concern about climate change and food security is that changing climatic conditions can initiate a vicious circle where infectious disease causes or compounds hunger, which, in turn, makes the affected populations more susceptible to infectious disease. The result can be a substantial decline in labour productivity and an increase in poverty and even mortality. Essentially all manifestations of climate change, be they drought, higher temperatures, or heavy rainfalls have an impact on the disease pressure, and there is growing evidence that these changes affect food safety and food security.

The recent IPCC report also emphasizes that increases in daily temperatures will increase the frequency of food poisoning, particularly in temperate regions. Warmer seas may contribute to increased cases of human shellfish and reef-fish poisoning (ciguatera) in tropical regions and a poleward expansion of the disease. However, there is little new evidence that climate change significantly alters the prevalence of these diseases. Several studies have confirmed and quantified the effects of temperature on common forms of food poisoning, such as salmonellosis. These studies show an approximately linear increase in reported cases for each degree increase in weekly temperature. Moreover, there is evidence that temperature variability affects the incidence of diarrhoeal disease. A number of studies found that rising temperatures were strongly associated with increased episodes of diarrhoeal disease in adults and children. These findings have been corroborated by analyses based on monthly temperature observations. Several studies report a strong correlation between monthly temperature and diarrhoeal episodes on the Pacific Islands, Australia, and Israel.

Extreme rainfall events can increase the risk of outbreaks of water-borne diseases particularly where traditional water management systems are insufficient to handle the new extremes. Likewise, the impacts of flooding will be felt most strongly in environmentally degraded areas, and where basic public infrastructure, including sanitation and hygiene, is lacking. This will raise the number of people exposed to water-borne diseases (*e.g.*, cholera) and thus lower their capacity to effectively use food.

Impacts of Climate Change on Access to Food

Access to food refers to the ability of individuals, communities, and countries to purchase sufficient quantities and qualities of food. Over the last 30 years, falling real prices for food and rising real incomes have led to substantial improvements in access to food in many developing countries. Increased purchasing power has allowed a growing number of people to purchase not only more food but also more nutritious food with more protein, micronutrients, and vitamins.

East Asia and to a lesser extent the Near-East/North African region have particularly benefited from a combination of lower real food prices and robust income growth. From 1970 to 2001, the prevalence of hunger in these regions, as measured by FAO's indicator of undernourishment, has declined from 24 per cent to 10.1 per cent and 44 per cent to 10.2 per cent respectively. In East Asia, it was endogenous income growth that provided the basis for the boost in demand for food, which was largely produced in the region; in the Near-East North African region demand was spurred by exogenous revenues from oil and gas exports, and additional food supply came largely from imports. But in both regions, improvements in access to food have been crucial in reducing hunger and malnutrition.

FAO's longer-term outlook to 2050 suggests that the importance of improved demand side conditions will even become more important over the next 50 years. The regions that will see the strongest reductions in the prevalence of undernourishment are those that are expected to see the highest rates of income growth. South Asia stands to benefit the most. Spurred by high income growth the region is expected to reduce the prevalence of undernourishment from >22 per cent currently to 12 per cent by 2015 and just 4 per cent by 2050. Progress is also expected for sub-Saharan Africa, but improvements will be less pronounced and are expected to set in later. Over the next 15 years, for instance, the prevalence of undernourishment will decline less than in other regions, from H-33 per cent to a still worrisome 21 per cent, as significant constraints (soil nutrients, water, infrastructure, etc.) will limit the ability to further increase food production locally, while continuing low levels of income rule out the option of importing food. In the long run, however, sub-Saharan Africa is expected to see a more substantial decline in hunger; by 2050, <6 per cent of its total population are expected to suffer from chronic hunger. It is important to note that these FAO projections do not take into account the effects of climate change.

By coupling agro-ecologic and economic models, others have gauged the impact of climate change on agricultural gross domestic product (GDP) and prices. At the global level, the impacts of climate change are likely to be very small; under a range of SRES and associated climate-change scenarios, the estimates range from a decline of "1.5 per cent to an increase of +2.6 per cent by 2080.

At the regional level, the importance of agriculture as a source of income can be much more important. In these regions, the economic output from agriculture itself will be an important contributor to food security. The strongest impact of climate change on the economic output of agriculture is expected for sub-Saharan Africa, which means that the poorest and already most food-insecure region is also expected to suffer the largest contraction of agricultural incomes. For the region, the losses in agricultural GDP, compared with no climate change, range from 2 per cent to 8 per cent for the Hadley Centre Coupled Model, version 3 and 7–9 per cent for the Commonwealth Scientific and Industrial Research Organisation projections.

Impacts on Food Prices

Essentially all SRES development paths describe a world of robust economic growth and rapidly shrinking importance of agriculture in the long run and thus a continuation of a trend that has been underway for decades in many developing regions. SRES scenarios describe a world where income growth will allow the largest part of the world's population to address possible local production shortfalls through imports and, at the same time, find ways to cope with safety and stability issues of food supplies.

It is also a world where real incomes rise more rapidly than real food prices, which suggests that the share of income spent on food should decline and that even high food prices are unlikely to create a major dent in the food expenditures of the poor. However, not all parts of the world perform equally well in the various development paths and not all development paths are equally benign for growth. Where income levels are low and shares of food expenditures are high, higher prices for food may still create or exacerbate a possible food security problem. There are a number of studies that have ventured to measure the likely impacts of climate change on food prices. The basic messages that emerge from these studies are: first, on average, prices for food are expected to rise moderately in line with moderate increases of temperature (until 2050); some studies even foresee a mild decline in real prices until 2050. Second, after 2050 and with further increases in temperatures, prices are expected to increase more substantially. In some studies and for some commodities (rice and sugar) prices are forecast to increase by as much as 80 per cent above their reference levels without climate change.

Third, price changes expected from the effects of global warming are, on average, much smaller than price changes from socio-economic development paths. For instance, the SRES A2 scenario would imply a price increase in real cereal prices by H-170 per cent. The (additional) price increase caused by climate change (in the Hadley Centre Coupled Model, version 3, climate change case) would only be 14.4 per cent.

Overall, this appears to be the sharpest price increase reported and it is not surprising that this scenario would imply a stubbornly high number of undernourished people until 2080. However, it is also needless to say that a constant absolute number of undernourished people would still imply a sharp decline in the prevalence of hunger, and, given the high population assumptions in the SRES A2 world (>13.6 billion people globally and >11.6 billion in the developing world) this would imply a particularly sharp drop in the prevalence from currently 17 per cent to H-7 per cent by 2080.

8

Microbiology

Microbiology is the study of microorganisms, which are microscopic, unicellular, and cell-cluster organisms. This includes eukaryotes such as fungi and protists, and prokaryotes. Viruses and prions, though not strictly classed as living organisms, are also studied. Microbiology typically includes the study of the immune system, or Immunology. Generally, immune systems interact with pathogenic microbes; these two disciplines often intersect which is why many colleges offer a paired degree such as "Microbiology and Immunology".

Microbiology is a broad term which includes virology, mycology, parasitology, bacteriology and other branches. A microbiologist is a specialist in microbiology and these other topics. Microbiology is researched actively, and the field is advancing continually. It is estimated only about one per cent of all of the microbe species on Earth have been studied. Although microbes were directly observed over three hundred years ago, the field of microbiology can be said to be in its infancy relative to older biological disciplines such as zoology and botany.

HISTORY

ANCIENT

The existence of microorganisms was hypothesized for many centuries before their actual discovery. The existence of unseen microbiological life was postulated by Jainism which is based on Mahavira's teachings as early as 6th century BCE.. Paul Dundas notes that Mahavira asserted existence of unseen microbiological creatures living in earth, water, air and fire. Jain scriptures also describe nigodas which are sub-microscopic creatures living in large clusters and having a very short life and are said to pervade each and every part of the universe, even in tissues of plants and flesh of animals.

The Roman Marcus Terentius Varro made references to microbes when he warned against locating a homestead in the vicinity of swamps "because there are bred certain minute creatures which cannot be seen by the eyes, which float in the air and enter the body through the mouth and nose and there cause serious diseases."

In 1546 Girolamo Fracastoro proposed that epidemic diseases were caused by transferable seedlike entities that could transmit infection by direct or indirect contact, or even without contact over long distances. However, early claims about the existence of microorganisms were speculative, and not based on any data or observation. Actual observation and discovery of microbes had to await the invention of the microscope in the 17th century.

MODERN

In 1676, Antonie van Leeuwenhoek observed bacteria and other microorganisms, using a single-lens microscope of his own design. While Van Leeuwenhoek is often cited as the first to observe microbes, Robert Hooke made the first recorded microscopic observation, of the fruiting bodies of molds, in 1665.. The first observation of microbes using a microscope is generally credited to the Dutch draper and haberdasher, Antonie van Leeuwenhoek, who lived for most of his life in Delft, Holland.

It has, however, been suggested that a Jesuit priest called Athanasius Kircher was the first to observe microorganisms. He was among the first to design magic lanterns for projection purposes, so he must have been well acquainted with the properties of lenses. One of his books, which reads in translation–'Concerning the wonderful structure of things in nature, investigated by Microscope.

Here, he wrote 'who would believe that vinegar and milk abound with an innumerable multitude of worms.' He also noted that putrid material is full of innumerable creeping animalculae. These observations antedate Robert Hooke's Micrographia by nearly 20 years and were published some 29 years before van Leeuwenhoek saw protozoa and 37 years before he described having seen bacteria. The field of bacteriology was founded in the 19th century by Ferdinand Cohn, a botanist whose studies on algae and photosynthetic bacteria led him to describe several bacteria including Bacillus and Beggiatoa. Cohn was also the first to formulate a scheme for the taxonomic classification of bacteria and discover spores.

Louis Pasteur and Robert Koch were contemporaries of Cohn's and are often considered to be the father of Microbiology and medical microbiology, respectively. Pasteur is most famous for his series of experiments designed to disprove the then widely held theory of spontaneous generation, thereby solidifying microbiology's identity as a biological science. Pasteur also designed methods for food preservation and vaccines against several diseases such as anthrax, fowl cholera and rabies.

Koch is best known for his contributions to the germ theory of disease, proving that specific diseases were caused by specific pathogenic microorganisms. He developed a series of criteria that have become known as the Koch's postulates. Koch was one of the first scientists to focus on the isolation of bacteria in pure culture resulting in his description of several novel bacteria including Mycobacterium tuberculosis, the causative agent of tuberculosis. While Pasteur and Koch are often considered the founders of microbiology, their work did not accurately reflect the true diversity of the microbial world because of their exclusive focus on microorganisms having direct medical relevance.

It was not until the late 19th century and the work of Martinus Beijerinck and Sergei Winogradsky, the founders of general microbiology, that the true breadth of microbiology was revealed. Beijerinck made two major contributions to microbiology: the discovery of viruses and the development of enrichment culture techniques.

While his work on the Tobacco Mosaic Virus established the basic principles of virology, it was his development of enrichment culturing that had the most immediate impact on microbiology by allowing for the cultivation of a wide range of microbes with wildly different physiologies. Winogradsky was the first to develop the concept of chemolithotrophy and to thereby reveal the essential role played by microorganisms in geochemical processes. He was responsible for the first isolation and description of both nitrifying and nitrogen-fixing bacteria.

THE SPECTRUM OF MICROBIOLOGY

Like all other living things, microorganisms are placed into a system of classification. Classification highlights characteristics that are common among certain groups while providing order to the variety of living things. The science of classification is known as taxonomy, and taxon is an alternative expression for a classification category. Taxonomy displays the unity and diversity among living things, including microorganisms. Among the first taxonomists was Carolus Linnaeus. In the 1750s and 1760s, Linnaeus classified all known plants and animals of that period and set down the rules for nomenclature.

CLASSIFICATION SCHEMES

The fundamental rank of the classification as set down by Linnaeus is the species. For organisms such as animals and plants, a species is defined as a population of individuals that breed among themselves. For microorganisms, a species is defined as a group of organisms that are 70 per cent similar from a biochemical standpoint.

In the classification scheme, various species are grouped together to form a genus. Among the bacteria, for example, the species *Shigella boydii* and *Shigella flexneri* are in the genus *Shigella* because the organisms are at least 70 per cent similar. Various genera are then grouped as a family because of similarities, and various families are placed together in an order. Continuing the classification scheme, a number of orders are grouped as a class, and several classes are categorized in a single phylum or division. The various phyla or divisions are placed in the broadest classification entry, the kingdom. Numerous criteria are used in establishing a species and in placing species together in broader classification categories. Morphology and structure are considered, as well as cellular features, biochemical properties, and genetic characteristics. In addition, the antibodies that an organism elicits in the human body are a defining property. The nutritional format is considered, as are staining characteristics.

PROKARYOTES AND EUKARYOTES

Because of their characteristics, microorganisms join all other living organisms in two major groups of organisms: prokaryotes and eukaryotes. Bacteria are prokaryotes because of their cellular properties, while other microorganisms such as fungi, protozoa, and unicellular algae are eukaryotes. Viruses are neither prokaryotes nor eukaryotes because of their simplicity and unique characteristics.

THE FIVE KINGDOMS

The generally accepted classification of living things was devised by Robert Whittaker of Cornell University in 1969. Whittaker suggested a five-kingdom classification. The first of the five kingdoms is Monera. Prokaryotes, such as bacteria and cyanobacteria are in this kingdom; the second kingdom, Protista, includes protozoa, unicellular algae, and slime molds, all of which are eukaryotes and single-celled; in the third kingdom, Fungi, are the molds, mushrooms, and yeasts. These organisms are eukaryotes that absorb simple nutrients from the soil. The remaining two kingdoms are Plantae and Animalia.

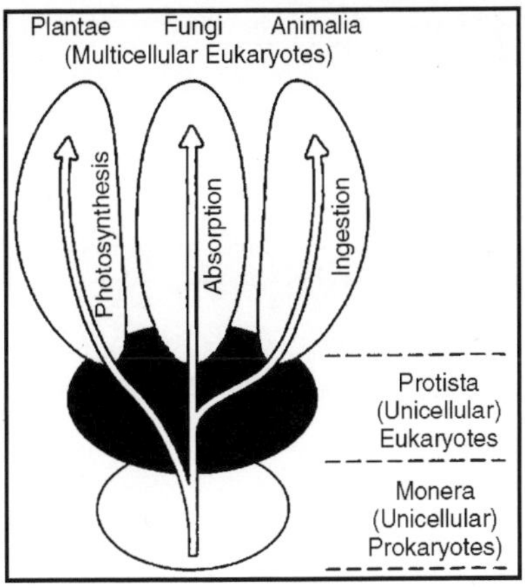

Fig. The Five Kingdoms of Living things Showing the Position of Microorganisms Relative to Other Organisms.

BRIEF DESCRIPTIONS OF MICROORGANISMS

Bacteria are relatively simple, prokaryotic organisms whose cells lack a nucleus or nuclear membrane. The bacteria may appear as rods, spheres, or spirals. Bacteria reproduce by binary fission, have unique constituents in their cell walls, and exist in most environments on earth. For instance, they live at temperatures ranging from 0° to 100°C and in conditions that are oxygen rich or oxygen free. A microscope is necessary to see and study them. Fungi are eukaryotic microorganisms that include multicellular molds and unicellular yeasts. The yeasts are slightly larger than bacteria and are used in alcoholic fermentations and bread making. Certain yeasts such as *Candida albicans* are pathogenic. Molds are filamentous, branched fungi that use spores for reproduction. The fungi prefer acidic environments, and most live at room temperature under oxygen-rich conditions. The common mushroom is a fungus.

Protozoa are eukaryotic, unicellular organisms. Motion is a characteristic associated with many species, and the protozoa can be classified just as to how they move: Some protozoa use flagella, others use cilia, and others use pseudopodia.

Certain species are nonmotile. Protozoa exist in an infinite variety of shapes because they have no cell walls. Many species cause such human diseases as malaria, sleeping sickness, dysentery, and toxoplasmosis. The term algae implies a variety of plantlike organisms. In microbiology, several types of single-celled algae are important. Examples are the diatoms and dinoflagellates that inhabit the oceans and are found at the bases of marine food chains. Most algae capture sunlight and transform it to the chemical energy of carbohydrates in the process of photosynthesis.

Viruses are ultramicroscopic bits of genetic material enclosed in a protein shell and, sometimes, a membranous envelope. Viruses have no metabolism; therefore, it is difficult to use drugs to interfere with their structures or activities. Viruses multiply in living cells and use the chemical machinery of the cells for their own purpose. Often, they destroy the cells in the process of replicating.

NOMENCLATURE OF MICROORGANISMS

The system for naming all living things, established by Linnaeus, is also applied to microorganisms. In this system, all organisms are placed into a classification system, and each organism is given a binomial name. The binomial name consists of two names. The first name is the genus to which the organism belongs.

The second name is a modifying adjective called the species modifier. In writing the binomial name, the first letter of the genus name is capitalized, and the remainder of the genus name and the complete species modifier are written in lowercase letters. The entire binomial name is either italicized or underlined. It can be abbreviated by using the first letter of the genus name and the full species modifier. An example of a microbial name is *Escherichia coli*, the bacterial rod found in the human intestine. The name is abbreviated *E. coli*.

MICROORGANISM

A microorganism or microbe is an organism that is unicellular or lives in a colony of cellular organisms. The study of microorganisms is called microbiology, a subject that began with Anton van Leeuwenhoek's discovery of microorganisms in 1675, using a microscope of his own design. Microorganisms are very diverse; they include bacteria, fungi, archaea, and protists; microscopic plants; and animals such as plankton and the planarian. Some microbiologists also include viruses, but others consider these as non-living.

Most microorganisms are unicellular but this is not universal, since some multicellular organisms are microscopic, while some unicellular protists and bacteria, like Thiomargarita namibiensis, are macroscopic and visible to the naked eye. Microorganisms live in all parts of the biosphere where there is liquid water, including soil, hot springs, on the ocean floor, high in the atmosphere and deep inside rocks within the Earth's crust.

Microorganisms are critical to nutrient recycling in ecosystems as they act as decomposers. As some microorganisms can fix nitrogen, they are a vital part of the nitrogen cycle, and recent studies indicate that airborne microbes may play a role in precipitation and weather. Microbes are also exploited by people in biotechnology, both in traditional food and beverage preparation, and in modern technologies based on genetic engineering. However, pathogenic microbes are harmful, since they invade and grow within other organisms, causing diseases that kill people, other animals and plants.

HISTORY

Evolution

Single-celled microorganisms were the first forms of life to develop on Earth, approximately 3–4 billion years ago. Further evolution was slow, and for about 3 billion years in the Precambrian eon, all organisms were microscopic. So, for most of the history of life on Earth the only forms of life were microorganisms. Bacteria, algae and fungi have been identified in amber that is 220 million years old, which shows that the morphology of microorganisms has changed little since the Triassic period.

Most microorganisms can reproduce rapidly and microbes such as bacteria can also freely exchange genes by conjugation, transformation and transduction between widely-divergent species. This horizontal gene transfer, coupled with a high mutation rate and many other means of genetic variation, allows microorganisms to swiftly evolve to survive in new environments and respond to environmental stresses.

This rapid evolution is important in medicine, as it has led to the recent development of 'super-bugs'—pathogenic bacteria that are resistant to modern antibiotics.

Pre-microbiology

The possibility that microorganisms exist was discussed for many centuries before their actual discovery in the 17th century. The existence of unseen microbiological life was postulated by Jainism which is based on Mahavira's teachings as early as 6th century BCE. Paul Dundas notes that Mahavira asserted existence of unseen microbiological creatures living in earth, water, air and fire.

Jain scriptures also describe nigodas which are sub-microscopic creatures living in large clusters and having a very short life and are said to pervade each and every part of universe, even in tissues of plants and flesh of animals. However, the earliest known idea to indicate the possibility of diseases spreading by yet unseen organisms was that of the Roman scholar Marcus Terentius Varro in a 1st century BC book titled On Agriculture in which he warns against locating a homestead near swamps:

- ...and because there are bred certain minute creatures which cannot be seen by the eyes, which float in the air and enter the body through the mouth and nose and there cause serious diseases.

In The Canon of Medicine, Avicenna stated that bodily secretion is contaminated by foul foreign earthly bodies before being infected. He also hypothesized that tuberculosis and other diseases might be contagious, *i.e.* that they were infectious diseases, and used quarantine to limit their spread. When the Black Death bubonic plague reached Andalusia in Spain, in the 14th century, Ibn Khatima wrote that infectious diseases were caused by contagious "minute bodies" that enter the human body.

Later, in 1546, Girolamo Fracastoro proposed that epidemic diseases were caused by transferable seedlike entities that could transmit infection by direct or indirect contact, or even without contact over long distances. All these early claims about the existence of microorganisms were speculative and were not based on any data or science. Microorganisms were neither proven, observed, nor correctly and accurately described until the 17th century. The reason for this was that all these early studies lacked the microscope.

History of Microorganisms' Discovery

Anton van Leeuwenhoek was one of the first people to observe microorganisms, using a microscope of his own design, and made one of the most important contributions to biology. Robert Hooke was the first to use a microscope to observe living things; his 1665 book Micrographia contained descriptions of plant cells.

Before Leeuwenhoek's discovery of microorganisms in 1675, it had been a mystery why grapes could be turned into wine, milk into cheese, or why food would spoil. Leeuwenhoek did not make the connection between these processes and microorganisms, but using a microscope, he did establish that there were forms of life that were not visible to the naked eye. Leeuwenhoek's discovery, along with subsequent observations by Lazzaro Spallanzani and Louis Pasteur, ended the long-held belief that life spontaneously appeared from non-living substances during the process of spoilage.

Lazzaro Spallanzani found that boiling broth would sterilise it and kill any microorganisms in it. He also found that new microorganisms could only settle in a broth if the broth was exposed to the air. Louis Pasteur expanded upon Spallanzani's findings by exposing boiled broths to the air, in vessels that contained a filter to prevent all particles from passing through to the growth medium, and also in vessels with no filter at all, with air being admitted via a curved tube that would not allow dust particles to come in contact with the broth.

By boiling the broth beforehand, Pasteur ensured that no microorganisms survived within the broths at the beginning of his experiment. Nothing grew in the broths in the course of Pasteur's experiment. This meant that the living organisms that grew in such broths came from outside, as spores on dust, rather than spontaneously generated within the broth. Thus, Pasteur dealt the death blow to the theory of spontaneous generation and supported germ theory.

In 1876, Robert Koch established that microbes can cause disease. He found that the blood of cattle who were infected with anthrax always had large numbers of Bacillus anthracis. Koch found that he could transmit anthrax from one animal to another by taking a small sample of blood from the infected animal and injecting it into a healthy one, and this caused the healthy animal to become sick. He also found that he could grow the bacteria in a nutrient broth, then inject it into a healthy animal, and cause illness.

Based on these experiments, he devised criteria for establishing a causal link between a microbe and a disease and these are now known as Koch's postulates. Although these postulates cannot be applied in all cases, they do retain historical importance to the development of scientific thought and are still being used today.

MICROBIAL ENZYMES

Indians are known to be lovers of milk and its products. As a sequel to white revolution, India has surged ahead to become the largest milk producer in the world, the production figure touched to 74 million tons for the year 1998. With growing urbanization, demand for processed dairy foods has increased considerably, in particular demand for different cheese varieties and low-lactose milk due to increasing intolerance of human beings to lactose in milk and other milk products. For improving the quality of milk and milk products, a number of different enzymes from microbial as well as from nonmicrobial sources have potential applications in dairy processing.

The use of rennet in cheese manufacture was among the earliest applications of exogenous enzymes in food processing, dating back to approximately 6000 B C. In 1994, the total production of cheese was 8000 metric tons against a total demand of 9000 metric tons. The projected demand by 2000 A D is around 30,000 metric tons. The use of rennet, as an exogenous enzyme, in cheese manufacture is perhaps the largest single application of enzymes in food processing.

In recent years, proteinases have found additional applications in dairy technology, for example in acceleration of cheese ripening, modification of functional properties and preparation of dietic products. Principal among some enzymes that have important and growing applications are lipases and b-galactosidases. Enzymes with limited applications include glucose oxidase, superoxide dismutase, sulphydryl oxidase, etc.

The increasing use of enzymes to produce specific products with characteristic attributes can be emphasized by the world-wide sale of industrial enzymes approximating to US $ 1.6 billion, which is expected to reach US $ 3.0 billion by the year 2008. Almost 45% of this is shared by the food industry and the remaining is shared by detergent (34.4%), textile (11%), leather (2.8%), pulp and paper (1.2%) industries and other industries (5.6%) excluding enzymes for use in diagnostics and therapeutics. India being the highest producer of milk in the world and consequently the surplus availability of milk in our country has triggered the food and dairy industry to convert the liquid milk into value-added products using biochemical and enzymatic processes.

MICROBIAL RENNETS

Animal rennet (bovine chymosin) is conventionally used as a milk-clotting agent in dairy industry for the manufacture of quality cheeses with good flavour and texture. Owing to an increase in demand for cheese production world wide – *i.e.* 4% per annum over the past 20 years, approximating 13.533 million tons – coupled with reduced supply of calf rennet, has therefore led to a search for rennet substitutes, such as microbial rennets. At present, microbial rennet is used for one-third of all the cheese produced world wide.

Rennin acts on the milk protein in two stages, by enzymatic and by nonenzymatic action, resulting in coagulation of milk. In the enzymatic phase, the resultant milk becomes a gel due to the influence of calcium ions and the temperature used in the process. Many microorganisms are known to produce rennet-like proteinases which can substitute the calf rennet.

Microorganisms like Rhizomucor pusillus, R. miehei, Endothia parasitica, Aspergillus oryzae and Irpex lactis are used extensively for rennet production in cheese manufacture. Extensive research that has been carried out so far on rennet substitutes has been reviewed by several authors. Different strains of species of Mucor are often used for the production of microbial rennets.

Whereas best yields of the milk-clotting protease from Rhizomucor pusillus are obtained from semisolid cultures containing 50% wheat bran, R.

miehei and Endothia parasitica are well suited for submerged cultivation. Using the former, good yields of milk-clotting protease may be obtained in a medium containing 4% potato starch, 3% soybean meal and 10% barley. During growth, lipase is secreted together with the protease. Therefore, the lipase activity has to be destroyed by reducing the pH, before the preparation can be used as cheese rennet.

Microbial rennets from various microorganisms (marketed under the trade names such as Rennilase, Fromase, Marzyme, Hanilase, etc.) being marketed since the 1970s have proved satisfactory for the production of different kinds of cheese. The molecular and enzymatic properties of chymosins have been studied extensively. Although the proteolytic specificities of the three commonly used fungal rennets are considerably different from those of calf chymosin, these rennets have been used to produce acceptable cheeses.

Recently Novo Nordisk has succeeded in expressing just one proteolytic enzyme from the fungus R. miehei in the well-known organism A. oryzae. This host organism is able to produce the single protease that cleaves the casein into a glycomacropeptide and para casein by hydrolysing only at the phe105–met106 peptide bond between phenyl alanine and methionine. This monocomponent enzyme product has the trade name Novoren. One major drawback of microbial rennet use in cheese manufacture, is the development of off flavour and bitter taste in the nonripened as well as in the ripened cheeses.

The rennets from microbial sources are more proteolytic in nature in comparison to rennet from animal sources, resulting in production of some bitter peptides during the process of cheese ripening. Hence, attempts have been made to clone the gene for calf chymosin and to express it in selected bacteria, yeasts and molds.

Recombinant Rennets

Due to shortage of calf stomachs and the economic value of cheese rennet, gene for calf chymosin was one of the first genes for mammalian enzymes that was cloned and expressed in microorganisms. Many different laboratories have cloned the gene for calf prochymosin in Escherichia coli and analysed the structure of the gene as well as the properties of the recombinant chymosin. The expressed proenzyme in E. coli, is present mainly as insoluble inclusion bodies, comprised of reduced prochymosin as well as molecules that are interlinked by disulphide bridges. After disintegration of the cells, inclusion bodies are harvested by centrifugation.

The individual laboratories have reported some differences in the procedure for renaturation of prochymosin from the inclusion bodies, but all have followed the same general scheme. The enzymatic properties of recombinant E. coli chymosin are indistinguishable from those of native calf chymosin. The enzymes were identical when observed by immunodiffusion in gels, but a slight difference was observed by enzyme linked immunosorbent assay (ELISA).

Table. Recombinal Chymosin Proparations at/or Approaching Legal and Commercial Acceptance.

Sourcc of DNA	Praducing microorganisms	Prducing company (brand namc)
Calf abomasums	klayveromyces lactis	Gist Brocades (Maxiren)
Calf abomasums	Aspergillus niger	Genccor/chr. Hansen (Chyrnoger)
Synthetic	Escherichic coli	Pfizer (Chymax)

The gene for prochymosin has also been cloned in Saccharomyces cerevisiae; the levels of expression have been reported to be 0.5 to 2.0% of total yeast protein. In yeast, about 20% of the prochymosin can be released in soluble form which can be activated directly; the remaining 80% is still associated with the cell debris.

The cloning of chymosin was carried out without the prosequence. Though the level of chymosin mRNA was similar to that of prochymosin mRNA, no milk-clotting activity was observed in clones containing the chymosin gene only. The results suggest that the prosequence is essential for correct folding of the polypeptide chain. The zymogen for the aspartic proteases from R. pusillus, also called mucor rennin, has likewise been cloned and expressed in yeast. Studies on its conversion to active form showed that secretion of R. pusillus protease from recombinant yeast was dependent on glycosylation of the enzyme. Compared to yeast, the filamentous fungi generally secrete larger quantities of proteins into the culture medium. Furthermore, filamentous fungi secrete the heterologous proteins with correct folding of the polypeptide chain and process correct pairing of sulphydryl groups.

The gene for R. miehei protease has been expressed in A. oryzae. Prochymosin has been expressed in Kluyveromyces lactis, A. nidulans, A. niger and Trichoderma reesei. In most of the cases, the reported yields of the model systems were about 10–40 mg of enzyme per litre of culture medium. However, 3.3 g of enzyme per litre has also been achieved. Yeast, Kluyveromyces lactis, has recently been used as an efficient host for the secretion of recombinant chymosin, which has led to the development of large-scale production process for chymosin. If produced on an industrial scale, the yields will perhaps be of the latter magnitude.

Chymosins which are approaching or have marketing and legal acceptance and generally regarded as safe (GRAS). Most of the companies produce recombinant rennet of cattle calf origin in different microbial hosts, however in India, the major source of milk is buffalo, showing a different composition from that of cow. The rennet from buffalo source has inherent compatibility for clotting buffalo milk. Therefore, National Dairy Research Institute, Karnal, has taken a lead in cloning gene of buffalo chymosin.

The buffalo chymosin cDNA has been cloned in E. coli and partial N-terminal sequencing of the purified buffalo chymosin indicates that it is highly homologous to the cattle chymosin. Several cheese making experiments have been carried out with recombinant chymosin and the general aspects of recombinant chymosin have been dealt with in a report. Since most of the rennet (> 90 %) added to cheese milk is lost in the whey, immobilization would considerably extend its catalytic life. Several rennets have been immobilized, but their efficiency as milk coagulants has been questioned.

So, there is a fairly general support for the view that immobilized enzymes cannot coagulate milk properly owing to inaccessability of the Phe–Met peptide bond of k-casein and that the apparent coagulating activity of immobilized rennets is due to leaching of the enzyme from the support. Different types of conventional cheeses have been successfully made by using recombinant rennet on an experimental or pilot scale.

No major differences have been detected between cheeses made with recombinant chymosins or natural enzymes, regarding cheese yield, texture, smell, taste and ripening. The recombinant chymosins are identical with calf rennet according to the report on biochemical and genetic evidences.

Dairy Industry

Lactose, the sugar found in milk and whey and its corresponding hydrolase, lactase or b -galactosidase, have been extensively researched during the past decade. This is because of the enzyme immobilization technique which has given new and interesting possibilities for the utilization of this sugar. Because of intestinal enzyme insufficiency, some individuals and even a population, show lactose intolerance and difficulty in consuming milk and dairy products.

Hence, low-lactose or lactose-free food aid programme is essential for lactose-intolerant people to prevent severe tissue dehydration, diarrhoea and, at times, even death. Another advantage of lactase-treated milk is the increased sweetness of the resultant milk, thereby avoiding the requirement for addition of sugars in the manufacture of flavoured milk drinks.

Manufacturers of ice cream, yoghurt and frozen desserts use lactase to improve scoop and creaminess, sweetness and digestibility and to reduce sandiness due to crystallization of lactose in concentrated preparations. Cheese manufactured from hydrolysed milk ripens more quickly than the cheese manufactured from normal milk.

Technologically, lactose crystallizes easily which sets limits to certain processes in the dairy industry and the use of lactase to overcome this problem has not reached its fullest potential because of the associated high costs. Moreover, the main problem associated with discharging large quantities of cheese whey is that it pollutes the environment. But, the discharged whey could be exploited as an alternate cheap source of lactose for the production of lactic acid by fermentation. The whey permeate, which is a by-product in the manufacture of whey protein concentrates, by ultrafiltration could be fermented efficiently by Lactobacillus bulgaricus.

Lactose can be obtained from various sources like plants, animal organs, bacteria, yeasts (intracellular enzyme), or molds. Some of these sources are used for commercial enzyme preparations. Lactase preparations from A. niger, A. oryzae and Kluyveromyces lactis are considered safe because these sources already have a history of safe use and have been subjected to numerous safety tests. The most investigated E. coli lactase is not used in food processing because of its cost and toxicity problems.

LACTASE

The properties of the enzyme depend on its source. Temperature and pH optima differ from source to source and with the type of particular commercial preparation. Immobilization of the enzymes, method of immobilization and type of carrier can also influence these optima values. In general, fungal lactase have pH optima in the acidic range 2.5–4.5 and yeast and bacterial lactases in the neutral region 6–7 and 6.5–7.5, respectively.

The variation in pH optima of lactases makes them suitable for specific applications, for example fungal lactases are used for acid whey hydrolysis, while yeast and bacterial lactases are suitable for milk (pH 6.6) and sweet whey (pH 6.1) hydrolysis. Product inhibition, e.g. inhibition by galactose, is another property which also depends on the source of lactase.

The enzyme from A. niger is more strongly inhibited by galactose than that from A. oryzae. This inhibition can be overcome by hydrolysing lactose at low concentrations by using immobilized enzyme systems or by recovering the enzyme using ultrafiltration after batch hydrolysis. Lactase from Bacillus species are superior with respect to thermostability, pH operation range, product inhibition and sensitivity against high-substrate concentration.

Thermostable enzymes, able to retain their activity at 60°C or above for prolonged periods, have two distinct advantages viz. they give higher conversion rate or shorter residence time for a given conversion rate and the process is less prone to microbial contamination due to higher operating temperature. Bacillus species have a pH optima of 6.8 and a temperature optima of 65°C. Its high activity for skim milk and less inhibition by galactose has made it suitable for use as a production organism for lactase.

The enzymatic hydrolysis of lactose can be achieved either by free enzymes, usually in batch fermentation process, or by immobilized enzymes or even by immobilized whole cells producing intracellular enzyme. Although numerous hydrolysis systems have been investigated, only few of them have been scaled up with success and even fewer have been applied at an industrial or semi-industrial level. Several acid hydrolysis systems have been developed to industry-scale level.

Large-scale systems which use free enzyme process have been developed for processing of UHT-milk and processing of whey, using K. lactis lactase (Maxilact, Lactozyme). Several commercial immobilized systems have been developed for commercial exploitation. Snamprogretti process of industrial-scale milk processing technology in Italy is one such working systems. They make use of fibre-entrapped yeast lactase in a batch process and the milk used is previously sterilized by UHT.

For pilot plants, there are three other processes designed and developed to handle milk:

- Gist-Brocades, Rohm GmbH (Germany);
- Sumitomo, Japan. These are continuous processes with short residence times. Processing of whey UF-permeate is accomplished by the system developed by Corning Glass, Connecticut, Lehigh, Valio and Amerace corp. The process by Corning Glass is being applied at commercial scale in the bakers yeast production using hydrolysed whey.

Microbial Enzymes

Cheese ripening is a complex process mediated by biochemical and biophysical changes during which a bland curd is developed into a mature cheese with characteristic flavour, texture and aroma.

The desirable attributes are produced by the partial and gradual breakdown of carbohydrates, lipids and proteins during ripening, mediated by several agents, viz:

- Residual coagulants,
- Starter bacteria and their enzymes,
- Nonstarter bacteria and their enzymes,
- Indegenous milk enzymes, especially proteinases;
- Secondary inocula with their enzymes.

Proteolysis occurs in all the cheese varieties and is a prerequisite for characteristic flavour development that can be regulated by proper use of the above agents. Cheese ripening is essentially an enzymatic process which can be accelerated by augmenting activity of the key enzymes. This has the advantage of initiating more specific action for flavour development compared to use of elevated temperatures that can result in accelerating undesirable nonspecific reactions and consequently off flavour development.

Enzymes may be added to develop specific flavours in cheeses, for example lipase addition for the development of Parmesan or Blue-type cheese flavours. Attempts to accelerate the multiple secondary flavour-forming-reactions, e.g. Strecker degradation, have been scarce. The pathways leading to the formation of flavour compounds are largely unknown and therefore the use of exogenous enzymes to accelerate ripening is mostly an empirical process. Different microbial enzymes used to accelerate cheese ripening.

Studies on the hydrolysis of whole casein and isolated casein components to observe the kinetics and specificity of aspartic proteases in rennin, pepsin and four microbial rennet substitutes indicated the considerable differences in the reaction velocity and the extent of hydrolysis on the rennet curd yield. The rennet enzymes were active even in the later phases of cheese-ripening.

PROTEINASES AND PEPTIDASES

Proteolysis is characteristic of most cheese varieties and is indispensable for good flavour and textural development. Proteinases used in cheese processing include:

- Plasmin,
- Rennet;
- Proteinases (cell wall and/or intracellular) of the starter and nonstarter bacteria. Approximately 6% of the rennet added to cheese milk remains in the curd after manufacture and contributes significantly to proteolysis during ripening. Combinations of individual neutral proteinases and microbial peptidases intensified cheese flavour and when used in combinations with microbial rennets reduced the intensity of bitterness caused by the latter. Acid proteases in isolation cause intense bitterness. Various animal or microbial lipases gave pronounced cheese flavour, low bitterness and strong rancidity, while lipases in combination with proteinases and/or peptidases give good cheese flavour with low levels of bitterness. In a more balanced approach to the acceleration of cheese ripening using mixtures of proteinases and peptidases, attenuated starter cells or cell-free extracts (CFE) are being favoured.

Table. Microbial Enzymes Used to Ucoclcrate Chocse Ripening

Enzyme	Source
Microbial lipases	Aspergillus niger
	A. oryzae
	Rhizomucor miehei
Lactasce	Streptococcus lactis
	Ktuyveromyces sp.
	Escherichia coli
Microbial serline	
proteinases	A. niger
Neutral pmroteinases	Bacillus subtilis A. oryzae

PROTEOLYTIC ENZYMES

The proteolytic system of lactic acid bacteria is essential for their growth in milk and contributes significantly to flavour development in fermented milk products. The proteolytic system is composed of proteinases which initially cleaves the milk protein to peptides; peptidases which cleave the peptides to small peptides and amino acids; and transport system responsible for cellular uptake of small peptides and amino acids.

Lactic acid bacteria have a complex proteolytic system capable of converting milk casein to the free amino acids and peptides necessary for their growth. These proteinases include extracellular proteinases, endopeptidases, aminopeptidases, tripeptidases and proline-specific peptidases, which are all serine proteases. Apart from lactic streptococcal proteinases, several other proteinases from nonlactostreptococcal origin have been reported.

There are also serine type of proteinases, e.g. proteinases from Lactobacillus acidophillus, L. plantarum, L. delbrueckii sp. bulgaricus, L. lactis and L. helveticus. Aminopeptidases are important for the development of flavour in fermented milk products, since they are capable of releasing single amino acid residues from oligopeptides formed by extracellular proteinase activity.

Other Dairy Enzymes

Other enzymes used for dairy food application include:

- Proteases to reduce allergic properties of cow milk products for infants;
- Lipases for development of lipolytic flavours in speciality cheeses.

The functional properties of milk proteins may be improved by limited proteolysis through the enzymatic modification of milk proteins. An acid-soluble casein, free of off flavour and suitable for incorporation into beverages and other acid foods, has been prepared by limited proteolysis. The antigenicity of casein is destroyed by proteolysis and the hydrolysate is suitable for use in milk-protein-based foods for infants allergic to cow milk.

Lipolysis makes an important contribution to swiss cheese flavours, due mainly to the lipolytic enzymes of the starter cultures. The characteristic peppery flavour of Blue cheese is due to short-chain fatty acids and methyl ketones. Most of the lipolysis in Blue cheese is catalysed by Penicillium roqueforti lipase, with a lesser contribution from indigenous milk lipase.

The NOVO process for production of enzyme modified cheese (EMC) uses medium-aged cheese which is emulsified, homogenized and pasteurized, after which 'palatase' (a lipase from R. miehei) is added, with or without a proteinase and the blend is ripened at a high temperature for one to four days. The mixture is reheated, a paste results which is suitable for inclusion in soups, dips, dressings, or snack foods. EMC technology has been developed to produce a range of characteristic cheese flavours and flavour intensities, for example swiss, blue, cheddar, provolo-nemor or romano, suitable for inclusion at low levels in many products.

The claims that exogenous enzymes are effective in accelerating ripening have not led to their wide-spread use, possibly due to their high cost, difficulties in distributing them uniformly in the curd and the possible danger of over-ripening the cheese. The other minor enzymes having limited applications in dairy processing include glucose oxidase, catalase, superoxide dismutase, sulphydryl oxidase, lactoperoxidase and lysozymes.

Glucose oxidase and catalase are often used together in selected foods for preservation. Superoxide dismutase is an antioxidant for foods and generates H_2O_2, but is more effective when catalase is present. Thermally induced generation of volatile sulphydryl groups is thought to be responsible for the cooked off-flavour in ultra high temperature (UHT) processed milk.

Use of sulphydryl oxidase under aseptic conditions can eliminate this defect. The natural inhibitory mechanism in raw milk is due to the presence of low levels of lactoperoxidase (LP), which can be activated by the external addition of traces of H_2O_2 and thiocyanate. It has been reported that the potential of LP-system and its activation enhances the keeping quality of milk. Cow milk can be provided with protective factors by the addition of lysozyme, making it suitable as an infant milk. Lysozyme acts as a preservative by reducing bacterial counts in milk without affecting the The scope of application of minor enzymes to milk and milk products has been recently reviewed.

The global market for the production of microbial enzymes for use in dairy-products manufacture is considerably large, but is being dominated only by a limited number of enzyme producers. In India the microbial dairy enzymes requirement has been very limited till now. However, with the advent of technological processes for the manufacture of different varieties of milk products, such as cheeses by the State Dairy Federations, Co-operatives and Private Dairy Product Manufacturers like Amul, Vijaya, Verka, Dynamix, Nestle, Smith Kline Beechem, etc., the markets for the sale of such products in megacities and towns has been slowly growing for the past two to three years.

Presently, many of these microbial enzymes, such as microbial rennets and other enzymes, are being imported. Hence, there is a scope for the production of enzymes such as microbial rennet, lactase, proteinases and lipases indigenously. In the near future, the requirement for these enzymes is bound to increase by leaps and bounds, basically due to requirement of value-added dairy products in the country.

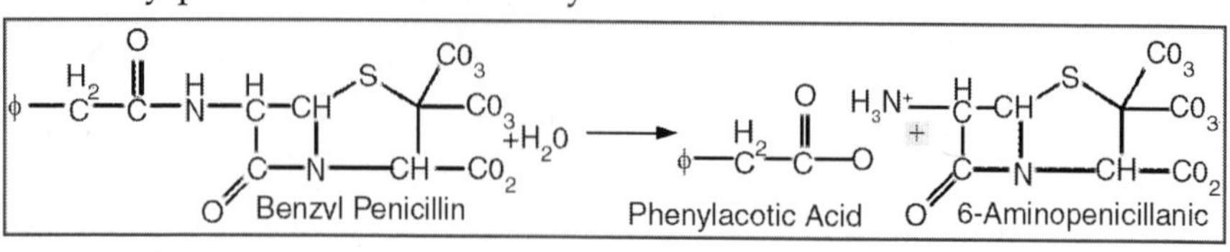

PRODUCTION OF ANTIBIOTICS

The mass production of antibiotics began during World War II with streptomycin and penicillin. Now most antibiotics are produced by staged fermentations in which strains of microorganisms producing high yields are grown under optimum conditions in nutrient media in fermentation tanks holding several thousand gallons.

The mold is strained out of the fermentation broth and then the antibiotic is removed from the broth by filtration, precipitation and other separation methods.

In some cases new antibiotics are laboratory synthesized, while many antibiotics are produced by chemically modifying natural substances; many such derivatives are more effective than the natural substances against infecting organisms or are better absorbed by the body, e.g., some semisynthetic penicillins are effective against bacteria resistant to the parent substance.

Benzylpenicillins and phenoxymethylpenicillins (penicillins G and V, respectively) are produced by fermentation and are the basic precursors of a wide range of semi-synthetic antibiotics, e.g. ampicillin. The amide link may be hydrolysed conventionally but the conditions necessary for its specific hydrolysis, whilst causing no hydrolysis of the intrinsically more labile but pharmacologically essential b-lactam ring, are difficult to attain. Such specific hydrolysis may be simply achieved by use of penicillin amidases (also called penicillin acylases).

Different enzyme preparations are generally used for the hydrolysis of the penicillins G and V, pencillin-V-amidase being much more specific than pencillin-G-amidase. Penicillin amidase may be obtained from E. coli and has been immobilised on a number of supports including cyanogen bromide-activated Sephadex G200.

It represents one of the earliest successful processes involving immobilised enzymes and is generally used in batch or semicontinuous STR processes (40,000 Ukgpenicillin G, 35°C, pH 7.8, 2 h) where it may be reused over 100 times. It has also been used in PBRs, where it has an active life of over 100 days, producing about two tonnes of 6-aminopenicillanic acid $kg^{-1}$of immobolised enzyme.

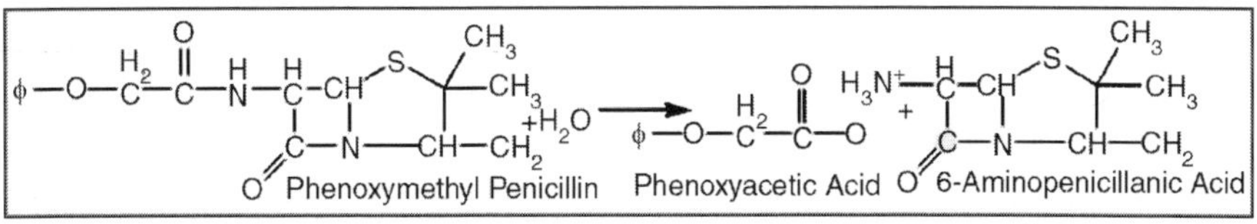

The penicillin-G-amidases may be used 'in reverse' to synthesise penicillin and cephalosporin antibiotics by non -equilibrium kinetically controlled reactions. Ampicillin has been produced by the use of penicillin-G-amidase immobilised by adsorption to DEAE -cellulose in a packed bed column:

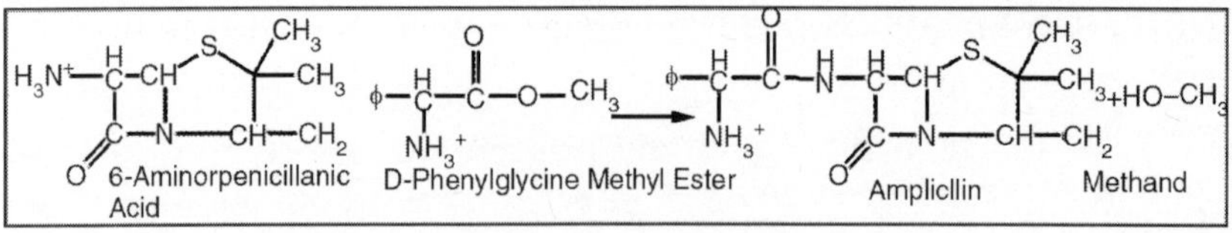

Many other potential and proven antibiotics have been synthesised in this manner, using a variety of synthetic b-lactams and activated carboxylic acids.

PREPARATION OF ACRYLAMIDE

Acrylamide is an important monomer needed for the production of a range of economically useful polymeric materials. It may be produced by the addition of water to acrylonitrile.

$$CH_2{=}CHCN + H_2O \rightarrow CH_2{=}CHCONH_2$$

This process may be achieved by the use of a reduced copper catalyst (Cu^+); however, the yield is poor, unwanted polymerisation or conversion to acrylic acid ($CH_2{=}CHCOOH$) may occur at the relatively high temperatures involved (80 -140°C) and the catalyst is difficult to regenerate. These problems may be overcome by the use of immobilised nitrile hydratase (often erroneously called a nitrilase). The enzyme from Rhodococcus has been used by the Nitto Chemical Industry Co. Ltd, as it contains only very low amidase activity which otherwise would produce unwanted acrylic acid from the acrylamide.

Immobilised nitrile hydratase is simply prepared by entrapping the intact cells in a cross-linked 10% (w/v) polyacrylamide/ dimethylaminoethylmethacrylate gel and granulating the product. It is used at 10°C and pH 8.0-8.5 in a semibatchwise process, keeping the substrate acrylonitrile concentration below 3% (w/v). Using 1% (w/v) immobilised-enzyme concentration (about 50,000 U l^{-1}) the process takes about a day.

Product concentrations of up to 20% (w/v) acrylamide have been achieved, containing negligible substrate and less than 0.02% (w/w) acrylic acid. Acrylamide production using this method is about 4000 tonnes per year. The closely related enzymes cyanidase and cyanide hydratase are used to remove cyanide from industrial waste and in the detoxification of feeds and foodstuffs containing amygdalin.

$$HCN + 2H_2O \rightarrow HCOO^- + NH_4^+$$

$$HCN + H_2O \rightarrow HCONH_2$$

IDENTIFYING USEFUL ANTIBIOTICS

Despite the wide variety of known antibiotics, less than 1% of antimicrobial agents have any medical or commercial value. The most commonly known antibiotic, Penicillin has a highly selective toxicity and therapeutic index (as eukaryotic animal cells do not contain peptidoglycan, they are usually unaffected by it). This is not so for many antibiotics.

Others simply lack advantage over the antibiotics already in use, or have no other practical applications.In order to identify the useful antibiotics, a process of screening is often employed. Using this method, isolates of a large number of microorganisms are cultured and then tested for production of diffusible products which inhibit the growth of test organisms.

However, the majority of the resulting antibiotics are already known and must therefore be disregarded. The remainders must be tested for their selective toxicities and therapeutic activities and the best candidates can be examined and possibly modified.A more modern version of this approach is a rational design programme. This involves screening being directed towards finding new natural products that inhibit specific targets (e.g. a particular step of a metabolic pathway) on microorganisms, rather than tests to show general inhibition of a culture.

Industrial Production Techniques

Antibiotics are produced industrially by a process of fermentation, where the source microorganism is grown in large containers (100,000–150,000 litres or more) containing a liquid growth medium. Oxygen concentration, temperature, pH and nutrient levels must be optimal and are closely monitored and adjusted if necessary.

As antibiotics are secondary metabolites, the population size must be controlled very carefully to ensure that maximum yield is obtained before the cells die. Once the process is complete, the antibiotic must be extracted and purified to a crystalline product. This is simpler to achieve if the antibiotic is soluble in organic solvent. Otherwise it must first be removed by ion exchange, adsorption or chemical precipitation.

Strains used for Production

Microorganisms used in fermentation are rarely identical to the wild type. This is because species are often genetically modified to yield the maximum amounts of antibiotics. Mutation is often used and is encouraged by introducing mutagens such as ultraviolet radiation, x-rays or certain chemicals. Selection and further reproduction of the higher yielding strains over many generations can raise yields by 20-fold or more.

Another technique used to increase yields is gene amplification, where copies of genes coding for proteins involved in the antibiotic production can be inserted back into a cell, via vectors such as plasmids. This process must be closely linked with retesting of antibiotic production and effectiveness.

TYPES OF ANTIBIOTICS

The great number of diverse antibiotics currently available can be classified in different ways, e.g., by their chemical structure, their microbial origin, or their mode of action. They are also frequently designated by their effective range. Tetracyclines, the most widely used broad-spectrum antibiotics, are effective against both Gram-positive and Gram-negative bacteria, as well as against rickettsias and psittacosis-causing organisms. Ciprofloxacin (Cipro) is another broad-spectrum antibiotic, effective in the treatment of mild infections of the urinary tract and sinuses.

The medium-spectrum antibiotics bacitracin, the erythromycins, penicillin and the cephalosporins are effective primarily against Gram-positive bacteria, although the streptomycin group is effective against some Gram-negative and Gram-positive bacteria. Polymixins are narrow-spectrum antibiotics effective against only a few species of bacteria.

Administration and Side Effects

Antibiotics are either injected, given orally, or applied to the skin in ointment form. Many, while potent anti-infective agents, also cause toxic side effects. Some, like penicillin, are highly allergenic and can cause skin rashes, shock and other manifestations of allergic sensitivity. Others, such as the tetracyclines, cause major changes in the intestinal bacterial population and can result in superinfection by fungi and other microorganisms. Chloramphenicol, which is now restricted in use, produces severe blood diseases and use of streptomycin can result in ear and kidney damage. Many antibiotics are less effective than formerly because antibiotic-resistant strains of microorganisms have emerged.

Non-medical Use

Antibiotics have found wide non-medical use. Some are used in animal husbandry, along with vitamin B_{12}, to enhance the weight gain of livestock. However, some authorities believe the addition of antibiotics to animal feeds is dangerous because continuous low exposure to the antibiotic can sensitize humans to the drug and make them unable to take the substance later for the treatment of infection. In addition low levels of antibiotics in animal feed encourage the emergence of antibiotic-resistant strains of microorganisms.

Drug resistance has been shown to be carried by a genetic particle transmissible from one strain of micro-organism to another and the presence of low levels of antibiotics can actually cause an increase in the number of such particles in the bacterial population and increase the probability that such particles will be transferred to pathogenic, or disease-causing, strains. Antibiotics have also been used to treat plant diseases such as bacteria-caused infections in tomatoes, potatoes and fruit trees. The substances are also used in experimental research.

MICROBES AND THE DAYS OF CREATION

The world of germs and microbes has received much attention in recent years—and for good reason. We frequently hear the term microbe associated with organisms such as Escherichia coli, Salmonella, anthrax bacteria, antibiotic-resistant tuberculosis, MRSA, HIV, malaria, Stachybotrys and other microscopic creatures. But where do microbes fit into the creation account? Were they created along with the rest of the plants and animals in the first week of creation, or were they created later, after the Fall?

Are microbes a result of the Curse? These and other questions are some that a group of professional creation microbiologists have been asking and their answers may surprise us. Ongoing research based on the creation paradigm appears to provide some answers to these puzzling questions.

Although we cannot be dogmatic about the details of microbe origin during Creation Week, it is believed that a reasonable extrapolation from biological data and Scripture can be made about the nature of microbes in a fully mature creation. Past creation scientists such as Leeuwenhoek, Pasteur and Lister, were blessed by God as He revealed to them critical insight into His creation. So, where do these microbes fit into the very good days of creation? Before answering this question, three terms, microbe, germ and symbiosis, need to be defined. These are relatively new or "modern" terms. First, the Bible does not use these specific terms. These terms were not commonly used until the end of the nineteenth century.

The term microbe was first used in 1878 to describe "extremely minute living beings." Before 1878, scientists including Louis Pasteur, used a variety of terms rather loosely to label the very small organisms that had interested them. It was not clear whether microbes belonged to the animal or plant kingdom, or to a completely different one. The term microbe was given by Charles E. Sedillot to describe bacteria Later, it would also be used of eukaryotic cells, including algae, fungi, protozoans and slime molds. Some people refer to viruses as microbes, but others do not because viruses are not cells. Viruses have nonliving, as well as living characteristics.

Today, the term germ refers to disease-causing microbes, or pathogens. All germs would have originated after the Fall The Edenic Curse would have profoundly influenced all creation, including viruses, bacteria, fungi and protozoans that would later become pathogens or parasites. The origin of infectious disease is complex and multifaceted. This topic is further explored in the book.

The Genesis of Germs It provides some understanding into the origin of infectious diseases. From a biblical worldview, infectious diseases and pathogenesis are a secondary state in nature. It is not the way the Creator intended for man and nature. Most microbes are beneficial to man and nature. Only about 5–10% of all bacteria are pathogenic. Many microbes live in a mutualistic relationship with plants, animals, humans and other microbes. Mutualism is a type of symbiosis. The term symbiosis is used to describe an intimate association between organisms of different species. Some symbiotic relationships between microbes and plants, animals and humans are essential for life on earth. For this reason most microbiologists maintain that bacteria, fungi, protists and other microbes have been maligned in the news media. Without our intestinal flora, we would not digest food nor acquire vitamins and minerals very efficiently. Without fungi, bacteria, algae and protozoans, life on earth could not last. This is because microbes provide essential "services" (e.g., nitrogen fixation, nitrification and denitrification) in nutrient cycles.

Viruses Fit into Creation

The determination of virus origin is uncertain. It may be that viruses have multiple origins. Some may be degenerate parts from cells after the Curse; still others may have their origin during the days of creation. Today, we think of viruses (Latin for "poison") only in the context of disease. However, some viruses (or at least virus-like genes) are involved in a positive function in nature.

Some groups of viruses, like bacteriophages, play a positive role in controlling bacteria in ecosystems and may play a role in diversity. Another group of viruses play a role in turning off the immune system during pregnancy in mammals and humans. This is a group referred to as endogenous retroviruses (ERVs). ERVs are among a kind of repetitious genetic elements called "retrotransposons". Research has shown that the ERV design prohibits the mother's immune system from damaging the child's body.

These retroviruses cannot fully replicate, only expressed in local immune cells (such as macrophages) of the placenta, thereby preventing them from initiating a full-blown immune response. Thus, the mother's immune system remains competent to respond to other infections but is specifically prevented from mounting an immune response to the developing embryo.

So in creation, the selective ability to turn off the immune system for protection would be a "good" design. Other ERVs also play a positive role in animal and human reproduction. However, since the corruption of creation, the corrupted retrovirus, HIV and various leukemia viruses turn off the entire immune system, leaving the body open to devastating infections.

These examples may provide clues to the origin of viruses and how some may have been created during Creation Week by design and how some have been corrupted as a result of the Fall.

Aseptic Technique

Trying to study this mixed population is often difficult and in the tradition of the scientific method; researchers dissect a system and study each piece in isolation. For microorganisms this means separating the organisms and getting them into pure culture. A pure culture is defined as a growth of microorganisms (a culture) that contains one cell type. It is essential in microbiology to be able to obtain and preserve pure cultures.

Over 100 years ago, Robert Koch devised methods to achieve this goal and the methods he developed are essentially still used today. The goals of aseptic technique are two-fold. The first objective is to obtain pure cultures and secondly to prevent cross-contamination. Microorganisms in culture must not escape into the environment and microbes in the environment must not get into the cultures we are studying. It is essential that aseptic technique be understood and practiced correctly.

Contaminated cultures are worthless for diagnosis or for doing research on, because it is unclear what microbe is performing any action that is being observed. Aseptic methods commonly used are flame sterilization, tube transfer, streak plates, spread plates and pour plates. Flame sterilization is an easy method to insure sterile transfer of a culture from a source to a growth medium.

Tube transfer is useful for moving inocula from one tube to another. Mechanical dilution by making streak plates is the preferred method for obtaining a pure culture of a microorganism. Finally, spread plates and pour plates are common methods for enumerating microorganisms and are sometimes useful for obtaining isolated colonies.

Flame Sterilization and Tube Transfer

Flame sterilization is a very quick simple method of killing microorganisms on an inoculating loop or needle. The loop or needle is held inside a flame for a few seconds to bring it to redness and then cooled. Once cool, thc loop or nccdlc can be used for various culture manipulations. Make sure that the area that contacts the culture is flamed to redness. Also, be patient and let the loop cool down, this usually takes about 15-30 seconds. Learning this technique is essential to everything else we do in microbiology.

Transfer of culture from agar plates to tubes, or from tube to tube, is a common, simple procedure. It is important to perform these transfers in a consistent and rapid manner. The following protocols have been found effective.

To transfer a culture from an agar plate to a broth or agar slant:

- Place the Bunsen burner in front of we and assemble all necessary equipment with in arms reach. Position everything so that we will not burn ourself while trying to inoculate our tubes.
- Label the tube of broth or agar to be inoculated with identifying marks. The culture, the date and our initials for example. Place it in a rack in front of we.
- Holding the inoculating loop handle, flame the entire wire to redness.
- When the wire cools (about 15-30 seconds) remove the lid of the plate with our other hand and obtain an inoculum by removing a small portion of the surface growth on the agar plate. In most cases we will be picking an isolated colony. Choose a well isolated one. Do not dig into the agar. Replace the lid of the plate immediately.

- Hold the tube to be inoculated with the free hand. Remove the cotton plug or cap of the tube with the little finger of the hand holding the needle holder. If a cotton-plugged tube is used, the mouth of the tube should be passed briefly through the flame to singe off dust and lint particles. (Dust or lint may fall into the tube and contaminate the medium.)
- Introduce the inoculum into the tube. When inoculating a tube of broth, rub the wire against the glass just above the fluid level and then tip the tube slightly to wash the inoculum into the broth. The wire should not be rattled against the sides of the tube to shake an inoculum into the broth; this is unnecessary and may create a dangerous aerosol. If the transfer is made to an agar slant, a single mid-line stroke over its surface is made with the wire or loop.
- Replace the cap or plug.
- Flame the inoculating wire again to redness, slowly to avoid spattering. Put the loop holder down after the wire cools.

Making a Medium

Making medium is as simple as cooking and a crude medium can be made in almost any kitchen with a few utensils and a source of heat. Below is described the production of a chicken broth medium that will grow many common microorganisms.

The medium will be sterilized by tyndaliza-tion. Simply boiling a medium once, while it will kill most vegetative cells, does not kill endospores.

However, heating encourages the endospores to germinate and a second boiling kills these microbes. The boiling process is repeated a third time to ensure that all spores have germinated and been killed.

- Add 250 ml (about 1 cup) of water into a glass container or some other vessel that can stand boiling water. The container should be something we can cover. Glass bottles that can stand boiling or canning jars work well.
- To this add 15 grams, about 1 tablespoon of instant chicken broth crystals. and stir until dissolved. Cover loosely so that steam can escape, but dust and dirt cannot enter.
- Microwave the broth on high for 3 minutes or until it just begins to boil.
- Let the broth cool and sit for at least 2 hours up to overnight at room temperature. Make sure that the medium vessel is covered to prevent contamination from the air. At this point the vegetative cells are dead and most endospores will germinate.
- Repeat steps 3 and 4 a second time. This will kill the endospores that germinated on day 1.
- Repeat steps 3 and 4 a third time. This will eliminate the remaining endospores, making the medium sterile.

- Place the medium in a warm place overnight. If our tyndalization was done correctly, our medium should remain clear and free of microbes.
- While we are waiting for our medium, search for a sample we would like to inoculate into it. Almost any sample should work, but natural samples are a good idea for this experiment. Samples coming from food or from our body may contain pathogens. Something we probably want to avoid.
- Check our medium. If it is still clear and sterile, add a pinch of our collected sample to the medium. (The exact amount really does not matter). Again place the medium in a warm place and check it periodically. After a few days, we should start to see the medium become more turbid.
- If we have the equipment, examine à sample of the medium under a microscope. What shapes of microbes are present? Is there more than one species?
- When finished, carefully dispose of the borth. It is basically a spoiled food, but the properties of the spoilage microbe are unknown and should be treated with care.

MICROBIAL CULTIVATION

When microorganisms are cultivated in the laboratory, a growth environment called a medium is used. The medium may be purely chemical (a chemically defined medium), or it may contain organic materials, or it may consist of living organisms such as fertilized eggs.Microorganisms growing in or on such a medium form a culture. A culture is considered a pure culture if only one type of organism is present and a mixed culture if populations of different organisms are present. When first used, the culture medium should be sterile, meaning that no form of life is present before inoculation with the microorganism.

General microbial media. For the cultivation of bacteria, a commonly used medium is nutrient broth, a liquid containing proteins, salts and growth enhancers that will support many bacteria. To solidify the medium, an agent such as agar is added. Agar is a polysaccharide that adds no nutrients to a medium, but merely solidifies it. The medium that results is nutrient agar. Many media for microorganisms are complex, reflecting the growth requirements of the microorganisms.

For instance, most fungi require extra carbohydrate and an acidic environment for optimal growth. The medium employed for these organisms is potato dextrose agar, also known as Sabouraud dextrose agar. For protozoa, liquid media are generally required and for rickettsiae and viruses, living tissue cells must be provided for best cultivation.

For anaerobic microorganisms, the atmosphere must be oxygen free. To eliminate the oxygen, the culture media can be placed within containers where carbon dioxide and hydrogen gas are generated and oxygen is removed from the atmosphere. Commercially available products achieve these conditions. Anaerobic chambers can also be used within closed compartments and technicians can manipulate culture media within these chambers. To encourage carbon dioxide formation, a candle can be burned to use up oxygen and replace it with carbon dioxide.

Special microbial media. Certain microorganisms are cultivated in selective media. These media retard the growth of unwanted organisms while encouraging the growth of the organisms desired. For example, mannitol salt agar is selective for staphylococci because most other bacteria cannot grow in its high-salt environment. Another selective medium is brilliant green agar, a medium that inhibits Gram-positive bacteria while permitting Gram-negative organisms such as *Salmonella* species to grow.

Still other culture media are differential media. These media provide environments in which different bacteria can be distinguished from one another.

For instance, violet red bile agar is used to distinguish coliform bacteria such as *Escherichia coli* from noncoliform organisms. The coliform bacteria appear as bright pink colonies in this media, while noncoliforms appear a light pink or clear. Certain media are both selective and differential. For instance, MacConkey agar differentiates lactose-fermenting bacteria from nonlactose-fermenting bacteria while inhibiting the growth of Gram-positive bacteria. Since lactose-fermenting bacteria are often involved in water pollution, they can be distinguished by adding samples of water to MacConkey agar and waiting for growth to appear.

In some cases, it is necessary to formulate an enriched medium. Such a medium provides specific nutrients that encourage selected species of microorganisms to flourish in a mixed sample. When attempting to isolate *Salmonella* species from fecal samples, for instance, it is helpful to place a sample of the material in an enriched medium to encourage *Salmonella* species to multiply before the isolation techniques begin.

In order to work with microorganisms in the laboratory, it is desirable to obtain them in pure cultures. Pure cultures of bacteria can be obtained by spreading bacteria out and permitting the individual cells to form masses of growth called colonies.

One can then pick a sample from the colony and be assured that it contains only one kind of bacteria. Cultivating these bacteria on a separate medium will yield a pure culture. To preserve microbial cultures, they may be placed in the refrigerator to slow down the metabolism taking place. Two other methods are deep-freezing and freeze-drying.

For deep-freezing, the microorganisms are placed in a liquid and frozen quickly at temperatures below –50°C. Freeze-drying (lyophilization) is performed in an apparatus that uses a vacuum to draw water off after the microbial suspension has been frozen. The culture resembles a powder and the microorganisms can be preserved for long periods in this condition.

STERILIZATION

Wet Heat

Wet heat is the most dependable procedure for the destruction of all forms of microbial life. Steam sterilization generally denotes heating in an autoclave employing saturated steam under a pressure of approximately 15 psi to achieve a chamber temperature of at least 121°C (250°F). The critical factors in insuring the reliability of this sterilization method is:

- Proper temperature and time; and
- The complete replacement of the air with steam (i.e. no entrapment of air).

Some autoclaves utilize a steam activated exhaust valve that remains open during the replacement of air by live steam until the steam triggers the valve to close. Others utilize a pre-cycle vacuum to remove air prior to steam introduction. Physical controls such as pressure gauges and thermometers are widely used but are considered secondary methods of insuring sterilization.

The use of appropriate biological indicators at locations throughout the autoclave is considered the best indicator of sterilization. The biological indicator most widely used for wet heat sterilization is Bacillus stearothermophilus spores.

Dry Heat

Dry heat is less efficient than wet heat sterilization and requires longer times and/or higher temperatures. The specific times and temperatures must be determined for each type of material being sterilized. Generous safety factors are usually added to allow for the variables that can influence the efficiency of this method of sterilization. The moisture of the sterilization environment as well as the moisture history of organisms prior to heat exposure appear to affect the efficiency of dry heat sterilization. Higher temperatures and shorter times may be used for heat resistant materials. The heat transfer properties and the spatial relation or arrangement of articles in the load are critical in insuring effective sterilization.

The advantage of wet heat is a better heat transfer to and into the cell resulting in overall shorter exposure time and lower temperature. Steam sterilization uses pressurized steam at 121-132° C (250-270° F) for 30 or 40 minutes. This type of heat kills all microbial cells including spores, which are normally heat resistant. In order to accomplish the same effect with dry heat in an oven, the temperature needs to be increased to 160-170° C (320-338° F) for periods of 2 to 4 hours.

General Procedures

All materials, equipment, apparatus contaminated with or containing potentially hazardous organisms should be steam sterilized before being washed and stored, or discarded. Autoclaving is the preferred method. Each individual working with biohazardous materials is responsible for sterilization of materials before disposal.

Biohazardous materials should not be placed in autoclaves overnight in anticipation of autoclaving the next day. To minimize hazard to firemen or disaster crews, all biohazardous materials must be placed in an appropriately marked refrigerator or incubator, sterilized, or otherwise confined at the close of each work day.

All autoclaves must be certified for operating efficiency by the periodic use of biological indicator controls and records maintained for three years. Special precautions should be taken to prevent accidental removal of material from an autoclave before it has been sterilized or the simultaneous opening of both doors on a double door autoclave. Dry hypochlorites, or any other strong oxidizing material, must not be autoclaved with organic materials such as a paper, cloth, or oil:

Oxidizer + Organic material + Heat = Possible Explosion.

All floors, laboratory benches and other surfaces in buildings where biohazardous materials are handled should be disinfected as often as deemed necessary by the supervisor. The surroundings should be disinfected after completion of operations involving plating, pipetting, centrifuging and similar procedures with biohazardous materials. It is the responsibility of the supervisor to determine that the disinfectant and the time and method of exposure is effective against the biological agent(s) used in the facility.

Floor drains should be flooded with water or disinfectant at least once each week in order to fill traps and thus prevent the back flow of sewer gases. Floor cleaning procedures which minimize the generation of aerosols should be used. Wet mopping or wet vacuum pick up is recommended.

Water used to mop floors should contain a disinfectant or disinfectant-detergent. (Dry mopping or dusting should be avoided). Where wet procedures are not practical, dry vacuum cleaning with a HEPA filter on the exhaust, sweeping compound used with push brooms, or dry dust mop heads treated to suppress aerosolization may be used. Stock solutions of suitable disinfectants will be maintained in each laboratory for disinfection purposes. NOTE: working dilutions of certain disinfectants have short shelf lives and must be prepared on a regular basis.

Sterilization Procedures

General criteria for sterilization of typical materials are presented below. Supervisors are encouraged to review the type of materials being handled and to establish standard conditions for sterilization. Treatment conditions to achieve sterility will vary in relation to the volume of material treated, volume of the autoclave, the contamination level, the moisture content and other factors.

Steam Autoclave:

- *Laundry:* 250°F (121°C) for a minimum of 30 min.
- *Trash:* 250°F (121°C) for at least 45 minutes per bag. Size of the autoclave and of the bags greatly effect sterilization time. Large bags in a small autoclave may require 90 minutes or more.
- *Glassware:* 250°F (121°C) for a minimum of 25 min.
- *Liquids:* 250°F (121°C) for 25 minutes for each gallon.
- *Animals & bedding:* Steam autoclaving not recommended (sterilization time required would be at least 8 hours). Incineration in an approved facility is the recommended treatment of these wastes.

Index

G

H

I

L

M

N

P

R

S

T

V

W